v *a* *d* *e* *m* *e* *c* *u* *m*

Urological Oncology

Daniel Nachtsheim, M.D.
Scripps Clinic
La Jolla, California, U.S.A.

LANDES
BIOSCIENCE
GEORGETOWN, TEXAS
U.S.A.

VADEMECUM
Urological Oncology
LANDES BIOSCIENCE
Georgetown, Texas U.S.A.

Please address all inquiries to the Publisher:
Landes Bioscience, 810 S. Church Street, Georgetown, Texas, U.S.A. 78626
Phone: 512/ 863 7762; FAX: 512/ 863 0081

ISBN: 1-57059-571-2

Library of Congress Cataloging-in-Publication Data

Urological oncology / [edited by] Daniel Nachtsheim.
     p. ; cm.
  Includes bibliographical references.
  ISBN 1-57059-571-2
  1. Genitourinary organs--Cancer.
  [DNLM: 1. Urologic Neoplasms. 2. Genital Neoplasms, Male--therapy. 3. Genital Neoplasms, Male. 4. Urologic Neoplasms--therapy. WJ 160 U78555 2004] I. Nachtsheim, Daniel.
  RC280.U74U775 2004
  616.99'46--dc22

                                2003015410

While the authors, editors, sponsor and publisher believe that drug selection and dosage and the specifications and usage of equipment and devices, as set forth in this book, are in accord with current recommendations and practice at the time of publication, they make no warranty, expressed or implied, with respect to material described in this book. In view of the ongoing research, equipment development, changes in governmental regulations and the rapid accumulation of information relating to the biomedical sciences, the reader is urged to carefully review and evaluate the information provided herein.

Contents

Editor

Daniel Nachtsheim, M.D.
Scripps Clinic
La Jolla, California, U.S.A.
DNachsheim@ScrippsClinic.com
Chapters 1

Contributors

Jonathan E. Bernie
University of California San Diego
San Diego, California, U.S.A.
Chapter 6

Christopher Cooper
Department of Urology
Children's Hospital of Iowa
Iowa City, Iowa, U.S.A.
Chapter 11

Daniel J. Cosgrove
University of California San Diego
San Diego, California, U.S.A.
Chapters 9, 10

Donald A. Elmajian
Louisiana State University
Shreveport, Louisiana, U.S.A.
Chapter 5

Huan Giap
Division of Radiation Oncology
Scripps Clinic
La Jolla, California, U.S.A.
Chapter 3

Karl Herwig
Scripps Clinic
La Jolla, California, U.S.A.
Chapter 8

Michael Kosty
Scripps Clinic
La Jolla, California, U.S.A.
Chapter 2

Marcus L. Quek
Department of Urology
University of Southern California
Los Angeles, California, U.S.A.
Chapter 7

Joseph D. Schmidt
University of California San Diego
San Diego, California, U.S.A.
Chapters 9, 10

Howard Snyder
Division of Pediatric Urology
Children's Hospital of Philadelphia
Philadelphia, Pennsylvania, U.S.A.
Chapter 11

John P. Stein
Department of Urology
University of Southern California
Los Angeles, California, U.S.A.
Chapter 7

Michael Tran
Division of Urology
Scripps Clinic
La Jolla, California, U.S.A.
Chapter 4

Pribhakar Tripuraneni
Scripps Clinic
La Jolla, California, U.S.A.
Chapter 3

Preface

This book devotes three chapters to the treatment of prostatic cancer, the most common tumor in men, and presents the three points of view for treatment of this complex issue. For localized disease, it has long been debated whether surgery or radiation is the best treatment. There are two good treatments, and the chapters outline which treatment may be suitable for the individual. For advanced disease, androgen deprivation has stood the test of time, and refinement of depriving prostate cancer of its growth factor, androgens, has been a major advance, though not curative. Recent research in advanced prostate cancer appears promising and will focus on non-hormonal factors such as gene therapy, prolongation of androgen sensitivity, and molecular triggers to induce apoptosis.

This book also reviews the major urological tumors and by and large supports the idea that early detection and surgical removal of the tumor results in long-term survival. The chapters on bladder and urinary collecting system tumors emphasize that superficial disease is common and may be treated by surgical excision endoscopically and treatment with topical chemotherapy and immune therapy such as BCG. Testis tumors demonstrate the success with surgical excision and chemotherapy, and what was considered a fatal tumor 25 years ago is now within the realm of cure in most cases. Chapter 7 on urinary diversions shows that recreation and reconstruction of the urinary system with bowel segments has many ramifications and can result in successful reconstruction without the need for ostomies.

Urological oncology is meant to be a quick read for the student or practitioner who needs an overall grasp of the organ system in question, and should give the reader a refined view of the problem and treatment in less than 1 hour reading per chapter.

Daniel Nachtsheim, M.D.
Scripps Clinic
La Jolla, California, U.S.A.

CHAPTER 1

Prostate Cancer: Local Disease

Daniel Nachtsheim

Lung and breast cancer are the most frequently diagnosed tumors in the United States. After lung cancer, prostate cancer is the second leading cause of death by cancer in men. Since the introduction of the prostate specific antigen (PSA) blood test, screening for prostate cancer has diagnosed over 180,00 new cases per year. Death from prostate cancer is on the decline, with 32,000 cases expected yearly. Over the past decade, mortality from prostate cancer has declined 1-2% per year. The impact of screening for prostate cancer also has resulted in migration of the initial diagnostic staging from advanced disease 20 years ago to the present where the majority of cases are diagnosed at a lower and localized stages. This is fortunate since metastatic prostate cancer is generally not curable by conventional therapy and over time will lead to death. Interventions with anti-androgens, luteinizing hormone releasing hormone (LHRH) or orchiectomy, however, may slow the progression of these androgen-dependent tumors and increase longevity. Over the past decade, treatment of localized prostate cancer by refined surgical and radiation therapy techniques has resulted in excellent survival rates at 5 and 10 years. In the following three chapters, we will discuss the treatment of localized disease by surgery and radiation techniques, and finally, the treatment of advanced disease.

Etiology and Clinical Factors

Etiology

Epidemiologic factors associated with the development of prostate cancer include age, genetic background, race, hormone status, and suspected dietary factors. Prostate cancer appears to be age-related occurring in 30% of men over age 50, and 80% of men over age 80, with an increase in incidence in each decade. It is important to note that many prostate cancers are considered incidental findings, are slow growing and do not lead to death. Recent studies have identified a form of prostate cancer that appears to be genetically linked to chromosome locus HPC1, and occurs in families at a somewhat earlier age, 40-60 years old. In a study of 14,000 men in The Netherlands, men who had one affected first-degree relative had a 1.41 relative risk of developing prostate cancer and with more than one affected relative the risk increased to 3.32, compared with that for men who did not have a positive family history. In addition, familial prostate cancer appears to be a more aggressive type of cancer at presentation.

Dietary factors have been linked to the development of prostate cancer. Several early studies indicated a high fat diet was linked to development of prostate cancer. However, subsequent studies have failed to find a direct relationship. Suggestive

Urological Oncology, edited by Daniel Nachtsheim. ©2005 Landes Bioscience.

1

evidence of dietary factors is most prominent in Asian countries where people consume low dietary fat and consume increased levels of soy-based foods. The beneficial effects of soy have been attributed to the isoflavones found in soy, genistein, and daidezine. Epidemiological evidence also suggests that selenium, an essential trace element, may play a role in human cancer prevention. In one study, a 50% decrease in total cancer mortality and a decrease in the incidence of prostate cancer occurred in a group of men that received 200 mcg daily of selenium versus a placebo.

Clinical Features

At presentation, 60% of patients have localized disease and only 15% have distant metastases at presentation since the PSA era. The majority of patients are either asymptomatic or have bladder outlet symptoms, such as urgency and frequency. Patients with metastatic disease most often complain of bone pain which can be in the low back, secondary to the distribution of veins into the lumbar area and a path for metastatic spread. Renal failure from ureter obstruction or anemia and low back pain may be initial signs of metastatic disease. Detection of prostate cancer is usually by digital rectal examination (DRE) and prostate specific antigen (PSA) testing. Transrectal ultrasound is insensitive and is rarely used for screening. Recommendations for PSA testing are somewhat controversial, but recommended by most urological organizations and should start at age 50 for an annual PSA examination and age 40 for those with a positive family history. PSA was identified in 1979 and came into usage in the mid-1980s. PSA is an antigen expressed exclusively in the prostatic epithelium and is a serine protease with a half-life of 2.2 days. PSA is expressed in normal prostate epithelium and is increased 10-fold per gram in prostate cancer cells. Despite being nonspecific it remains a valuable screening tool for the detection of prostate cancer. Other causes of PSA elevation include prostate infections, benign prostatic hyperplasia (BPH) and after vigorous digital rectal examination or immediately following ejaculation. PSA is elevated in 80-90% of men with prostate cancer. However, cancer can occur 10-20% of the time with normal levels of PSA. Recently, the fractionated PSA and bound PSA tests have been used to differentiate elevations of PSA from malignant versus benign processes. The percentage of PSA unbound to protein is measured and levels beyond 25% free PSA are associated with a low incidence of prostate cancer approaching the normal population. Medications can suppress PSA levels resulting in false-negative tests. Finasteride, an 5-alpha-reductase inhibitor used for treatment of BPH and baldness, will lower PSA by an average of 50% in a benign prostate and a lower percentage in prostate cancer. Also, patients who have had surgical or medical castration will have dramatically lower PSA values, and herbal medications frequently taken in the population may also lower PSA slightly. Herbal compounds containing phytoestrogens such as PC Spes have been found to lower PSA also.

Diagnosis and Treatment

Diagnosis of Prostate Cancer

Diagnosis of prostate cancer is most often made by needle biopsy of the prostate, following suspicion of its presence by digital rectal examination and PSA testing. Digital rectal examination may strongly suggest the diagnosis; however, tissue confirmation is needed and may proceed by transrectal or transperineal biopsy of the prostate under a local anesthetic or transurethral resection of the prostate. Since most cancers present in the peripheral zone of the prostate, transurethral resection is less likely to find cancer since it samples the central and transition zones of the prostate.

Histologic examination of the biopsy material is categorized by the Gleason method with a possible total score of 10 based on the major histological pattern combined with the minor histologic pattern. Gleason scores of 2, 3, and 4 are considered well-differentiated, 5 and 6 moderately differentiated, and 7, 8, 9, and 10 aggressive tumors with worse prognosis. An overall assessment of patient prognosis can be established by the level of PSA test, the Gleason score at presentation, and the volume of tumor estimated in the gland, as well as the clinical stage.

"Parten" tables are nomograms for predicting the probability of organ confined disease, capsular penetration of the prostate by tumor, seminal vesicle involvement and lymph node involvement based on levels of PSA and Gleason score, and can be useful in evaluating for patients for the type of treatment. Imaging studies may include a radionucleotide bone scan if the PSA test is more than 20, and can also include MRI scanning of the prostate and pelvic CT scanning for assessment of the extent of disease. Staging systems for prostate cancer have historically been based on the Whitmore-Jewett system of A, B, C, and D; however, in recent years conformity to the tumor node metastases, TNM (tumor, nodes, metastasis) system has prevailed and been adopted by the American Joint Committee for Cancer Staging. The systems are compatible (Fig. 1.1).

Treatment

While patients may have a difficult time accepting the diagnosis of prostate cancer, the decision regarding treatment is often the most difficult part of the process. Treatment for cure for localized disease generally involves whether to have surgical extirpation or radiation therapy. With well-differentiated tumors, a low PSA test in advancing age, watchful waiting is in favor, and may be acceptable for those willing to undergo repeated PSA tests and examination.

Prostate Surgery

Prostatectomy for Cure

Selection of candidates suitable for surgery is based on the natural history of prostate cancer, its Gleason score, the life expectancy of the candidate, and consideration of morbidity from the operation. The best candidates for total prostatectomy are those likely to benefit from it, and therefore, they should be young enough to enjoy the benefit from cure. Candidates ideally are those less than 75 years old in clinical stage T1B, T2A, or T2B, and some T2C. Some candidates with PSA cancers, that is T1C may also benefit if they have medium to high-grade Gleason scores.

Prostatectomy for cure in prostate cancer has been referred to as radical prostatectomy based on historic operation of removal of pelvic lymph nodes and wide excision of the prostate. More modern terminology would describe the operation as a total prostatectomy removing the prostate and seminal vesicles in a controlled fashion preserving the neural vascular bundles when possible, which lie next to the prostate, and preservation of the anatomical urinary sphincter. Excision of the pelvic lymph nodes can be omitted with low levels of PSA testing and well-differentiated tumors that have less than a 10% chance of lymph node involvement.

Prostate Cancer

Staging

Once cancer of the prostate has been found (diagnosed), more tests will be done to find out if cancer cells have spread from the prostrate to tissues around it or to other parts of the body. This is called "staging." To plan treatment, a doctor needs to know the stage of the disease. The following stages are used for cancer of the prostate.

Stage I (A)

Prostate cancer at this stage cannot be felt and causes no symptoms. The cancer is only in the prostate and usually is found accidentally when surgery is done for other reasons, such as for benign prostatic hyperplasia. Cancer cells may be found in only one area of the prostate or they may be found in many areas of the prostate.

Stage II (B)

The tumor may be found by a needle biopsy that is done because a blood test (called a prostate-specific antigen [PSA] showed an elevated PSA level or it may be felt in the prostate during a rectal examination, even though the cancer cells are found only in the prostate gland.

Stage III (C)

Cancer cells have spread outside the covering (capsule) of the prostate to tissues around the prostate. The glands that produce semen (the seminal vesicles) may have cancer in them.

Stage IV (D)

Cancer cells have spread (metastasized) to lymph nodes (near or far from the prostate) or to organs and tissues far away from the prostate such as the bone, liver, or lungs.

Recurrent

Recurrent disease means that the cancer has come back (recurred) after it has been treated. It may come back in the prostate or in another part of the body.

Prostate staging can also be described by using T (tumor size), N (extent of spread to lymph nodes), and M (extent of spread to other parts of the body).

TNM (Tumor, Node, Metastases)

Primary Tumor

TX   Primary tumor cannot be assessed

T0   No evidence of primary tumor

T1   Tumor not palpable nor visible by imaging

T1a Tumor incidental histologic finding in 5% or less of tissue resected

T1b Tumor incidental histologic finding in more than 5% of tissue resected

T1c Tumor identified by needle biopsy (eg, because of elevated PSA)

T2   Tumor confined within prostate*

T2a Tumor involves one lobe

T2b Tumor involves both lobes

T3   Tumor extends through capsule**

T3a Tumor extends on one or both sides

T3b Tumor invades seminal vesicle

T4   Tumor is fixed or invades adjacent structures other than seminal vesicles: bladder neck, external sphincter, rectum, levator muscles, and/or pelvic wall

Regional Lymph Nodes (N)

NX   Regional lymph nodes cannot be assessed

N0   No regional lymph node metastasis

N1   Metastasis in regional lymph node(s)

Metastases (M)

MX   Distant metastasis cannot be assessed

M0   No distant metastasis

M1   Distant metastasis

M1a Nonregional lymph node(s)

M1b Bone(s)

M1c Other sites(s)

*Note: Tumor found in one or both lobes by needle biopsy, but not palpable or reliably visible by imaging, is classified as T1c.

** Note: Invasion into the prostatic apex or into (but not beyond the prostatic capsule is not classified as T3, but as T2.

Fig. 1.1. Prostate cancer staging. Reprinted from: Walsh PC, Worthington JF. The Prostate: A Guide for Men and the Women Who Love Them. pp. 104-105. ©1995, with permission of The Johns Hopkins University Press.

Table 1.1. Partin—nomogram for prediction of final pathological stage

PSA 0.0–4.0 ng/ml — Clinical Stage

Score	T1c	T1b	T1c	T2a	T2b	T2c	T3a
Prediction of organ-confined disease							
2-4	100	85	92	88	76	82	–
5	100	78	81	81	67	73	–
6	100	68	69	72	54	60	42
7	–	54	55	61	41	46	–
8-10	–	–	–	48	31	–	–
Prediction of established capsular penetration							
2-4	0	15	22	14	26	17	–
5	0	22	30	20	34	26	–
6	0	30	34	29	46	38	59
7	–	43	40	39	59	50	–
8-10	–	–	–	50	68	–	–
Prediction of seminal vesicle involvement							
2-4	0	1	<1	1	2	2	–
5	0	3	<1	2	4	4	–
6	0	6	1	5	9	9	8
7	–	12	4	9	17	17	–
8-10	–	–	–	17	29	–	–
Prediction of lymph nodal involvement							
2-4	0	2	<1	1	2	4	–
5	0	3	<1	1	2	4	–
6	0	8	2	3	9	17	15
7	–	15	2	7	18	31	–
8-10	–	–	–	13	32	–	–

PSA 4.1–10 ng/ml — Clinical Stage

Score	T1c	T1b	T1c	T2a	T2b	T2c	T3a
Prediction of organ-confined disease							
2-4	100	78	82	83	67	71	–
5	100	70	71	73	56	64	43
6	100	53	59	62	44	48	33
7	100	39	43	51	32	37	26
8-10	–	32	31	39	22	25	12
Prediction of established capsular penetration							
2-4	0	22	29	19	34	27	–
5	0	29	34	28	45	34	58
6	0	45	38	38	56	49	68
7	0	58	44	49	68	59	75
8-10	–	64	48	59	77	71	87
Prediction of seminal vesicle involvement							
2-4	0	2	<1	1	3	3	–
5	0	4	3	3	6	5	5
6	0	9	1	6	11	12	11
7	0	18	5	12	22	23	18
8-10	–	29	23	22	38	40	40
Prediction of lymph nodal involvement							
2-4	0	2	1	1	2	5	–
5	0	2	1	1	2	5	–
6	0	9	2	4	11	19	16
7	0	18	5	8	20	34	28
8-10	–	30	5	15	35	53	50

PSA 10.1–20 ng/ml — Clinical Stage

Score	T1c	T1b	T1c	T2a	T2b	T2c	T3a
Prediction of organ-confined disease							
2-4	100	–	–	61	52	–	–
5	100	49	55	58	43	52	26
6	–	36	41	44	28	37	19
7	–	24	24	36	19	24	14
8-10	–	11	11	29	14	15	9
Prediction of established capsular penetration							
2-4	0	–	–	40	49	–	–
5	0	49	40	43	58	61	75
6	–	62	45	56	73	59	82
7	–	73	52	64	81	73	86
8-10	–	87	–	70	86	82	92
Prediction of seminal vesicle involvement							
2-4	0	–	–	3	4	–	–
5	0	7	<1	5	8	12	11
6	–	15	1	11	19	17	18
7	–	28	6	19	33	33	31
8-10	–	55	–	29	50	53	49
Prediction of lymph nodal involvement							
2-4	0	–	–	1	3	–	–
5	0	–	1	3	3	–	–
6	–	11	4	5	13	22	20
7	–	21	7	9	24	39	35
8-10	–	41	–	17	40	59	54

PSA Greater Than 20 ng/ml — Clinical Stage

Score	T1c	T1b	T1c	T2a	T2b	T2c	T3a
Prediction of organ-confined disease							
2-4	–	33	20	–	7	–	–
5	–	24	32	–	11	3	5
6	–	22	14	–	4	4	3
7	–	7	18	–	4	5	3
8-10	–	3	3	–	1	2	2
Prediction of established capsular penetration							
2-4	–	50	80	94	–	–	–
5	–	54	68	–	97	–	95
6	–	53	86	90	96	96	95
7	–	67	80	96	95	95	98
8-10	–	74	97	99	99	97	98
Prediction of seminal vesicle involvement							
2-4	–	<1	12	30	–	–	–
5	–	<1	11	–	29	–	31
6	–	2	35	40	53	62	55
7	–	9	31	73	73	73	65
8-10	–	81	81	93	–	–	–
Prediction of lymph nodal involvement							
2-4	–	6	2	7	–	–	–
5	–	6	2	7	–	–	–
6	–	8	9	18	53	62	3
7	–	24	11	44	62	73	55
8-10	–	41	35	76	73	65	–

Numbers represent the percent probability of the patient having a given final pathological stage based on a logistic regression analysis for all 3 variables combined. Dash represents lack of sufficient data to calculate probability.

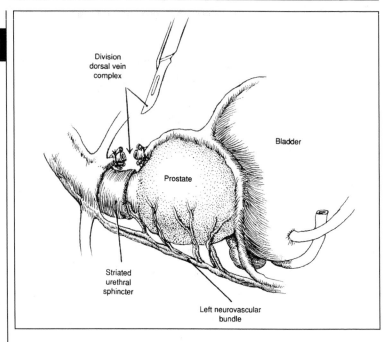

Fig. 1.2. Total prostatectomy.

Removal of the prostate is most often done by the retroperitoneal approach (Fig. 1.2), with an incision in the midline from the umbilicus to the pubis, palpation of the pelvic lymph nodes, and removal of the prostate by incising the endopelvic fascia on both sides of the prostate, taking down the puboprostatic ligaments, and dividing the urethra at the apex of the prostate and retrograde removal of the fascia along side of the prostate, taking care to avoid the neural vascular bundles responsible for erection. The prostate is circumferentially removed from the bladder neck, taking the seminal vesicles with the specimen (Fig. 1.3). The bladder neck is reconstituted with anastomosis to the urethra giving urinary continuity and ideally preserving urinary continence and potency.

Total prostatectomy may also be achieved by the perineal route, making an incision across the space between the scrotum and the anus, giving immediate access to the prostate. With careful dissection, the neural vascular bundles can also be spared, and the prostate removed in a similar fashion to the retroperitoneal approach. Lymph node dissection is not possible by this method unless done separately by open incision or laparoscopic means. The perineal approach has been favored by some surgeons as giving better continuity of the urethra to the bladder. Complications of total prostatectomy include total incontinence in 3-5% of patients, stress incontinence in 9%, and erectile impotence from 30-50%. Walsh's modifications of prostatectomy include the nerve-sparing prostatectomy, which has allowed preservation of potency in 50-70% of men in the younger age groups.

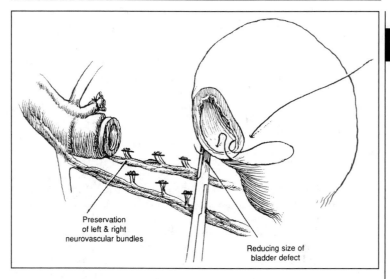

Fig. 1.3. Prostatectomy.

Cryoablation of the prostate which involves freezing of the prostate, has been reintroduced as a form of treatment for prostate cancer, which is localized. There are no long-term studies to show how well this works, but in the early experience many men have a positive biopsy following the treatment and a high incidence of complications secondary to rectal injury and fistula. It remains to be determined whether this form of treatment will be as effective as external beam radiation or total prostatectomy in curing prostate cancer. Cryosurgery has also been used following failure of radiation therapy to the prostate.

Transurethral resection of the prostate, removing the inner core of the prostate, but leaving the capsule intact, remains an option for some men who have bulky obstruction, which may be tumor or mixture of BPH and tumor. Older men who need relief of obstructive symptoms and not necessarily cure, may undergo transurethral resection of the prostate, which may be combined by hormone ablation by medication or orchiectomy.

Treatment of complications from prostatectomy is usually successful. Urinary incontinence of a mild nature can be improved by anticholinergic medication such as oxybutynin or tolterodine tartrate to relax the bladder, strengthening of the urinary sphincter with muscle exercises, and occasionally with injection of the vesicle neck transurethrally with collagen bulking agents. Ultimately, total incontinence can be cured with implantation of the artificial urinary sphincter. Treatment for erectile impotence may be quite successful with penile injection of vasodilators such as prostaglandin or a triple mixture of papaverine, prostaglandin, and phentolamine. The vacuum ErecAid device, an external vacuum pump with compression band, is also useful, and sildenafil (Viagra) 50-100 mg tablets has also been used successfully following prostatectomy. Ultimately, insertion of a semi-rigid or inflatable penile prosthesis can be used for complete rehabilitation, as patients remain capable of achieving orgasm following surgery.

Survival

Since prostate cancer is a slow growing tumor, it is pertinent to look at survival after treatment at the 5- to 15-year period. Most data for survival appears favorable at the 5-year mark, but declines before 10 years. In general, about 70% of men treated by radical prostatectomy are cured. Approximately 20% of men will have PSA recurrence, chemical or x-ray evidence of cancer recurrence most likely within 3 years of surgery, but can take as long as 10 years. Selective radiation therapy or systemic hormone therapy can be used in this situation.

In one contemporary series the 7-year all-cause survival was 90% and the cancer-specific survival rate was 97% (Catalona). This is consistent with the Scripps Clinic cancer registry data which indicate an all-cause survival of 85-90% at 7 years following surgery (Fig. 1.4). These results are higher than historical surgical data, as well as external beam radiation and conservative management.

In a study by the National Cancer Institute of 59,576 patients, the 10-year survival rate for prostate cancer was used as the endpoint. The results showed that for Gleason scores 2-4, the 10-year survival was 98% for those treated with surgery, 89% for radiotherapy, and 92% for those managed conservatively.

For those with moderately differentiated tumors, Gleason 5-7, the rates were 91% with surgery, 74% for radiotherapy, and 76% conservative management. For aggressive tumors, Gleason grade 8-10, the results were 76%, 52%, and 43%, respectively. These results imply that patients with moderate to poorly differential tumors fare better with prostatectomy.

Presently, more men are being diagnosed with prostate cancer at an earlier age and stage. The proportion of organ confined disease has doubled. Certainly the optimum treatment in terms of survival, freedom from disease and quality of life needs ongoing analysis.

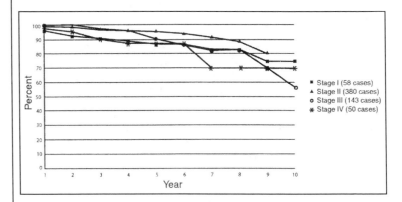

Fig. 1.4. Scripps Clinic. 1987-1996 Prostate: Cases treated with primary prostatectomy. 10-year observed Kaplan-Meier survival analysis (n=643 analytic cases)

Selected Readings

1. Section XI: Prostate cancer. In: Campbell's Urology Textbook, 7th ed. W.B. Saunders Co., 1998:2489-2658.
2. Stamey TA, Yemoto CM, McNeal JE et al. Prostate cancer is highly predictable: A prognostic equation based on all morphological variables in radical prostatectomy specimens. J Urol 2000; 163:1155-1160.
3. Pound CR, Partin AW, Eisenberger MA et al. Natural history of progression after PSA elevation following radical prostatectomy. JAMA 1999; 281:1591-1597.
4. Catalona WJ, Smith DS. Cancer recurrence and survival rates after anatomic radical retropubic prostatectomy for prostate cancer: Intermediate term results. J Urol 1998; 160:2428-2434.
5. Lu-Yao GL, Yao S-L. Population-based study of long-term survival in patients with clinically localized prostate cancer. Lancet 1997; 349:906-910.
6. Walsh PC. The Prostate. The Johns Hopkins University Press, 1995.
7. Walsh PC, Partin AW, Epstein JI. Cancer control and quality of life following anatomical radical retropubic prostatectomy: Results at 10 years. J Urol 1994; 152:1831-1836.

1

CHAPTER 2

Treatment of Advanced Prostate Cancer

Michael Kosty

Despite recent advances in treatment of localized prostate cancer and promising data for regional disease, 88 men will die of prostate cancer each day in 2003. With increased use of prostate-specific antigen (PSA) as a screening tool and directed biopsy techniques, more patients are being diagnosed at an early, potentially curable stage. Only 15% of patients present with metastatic disease, down from approximately 40% in 1980. Over the past five years, the number of individuals dying of prostate cancer has declined from nearly 40,000 to 31,900 in 2000. This may be related to the increased use of PSA as a screening tool, allowing patients to be diagnosed at an earlier, potentially curable stage. For metastatic disease, hormonal manipulation remains the initial therapy of choice. However, once a patient progresses from initial androgen deprivation and develops symptomatic metastasis, the median survival is approximately 10 months. Recent advances in the treatment of hormone refractory disease offer patients palliation of symptomatic disease and may lead to improved overall survival.

Androgen Deprivation

There are five methods of androgen deprivation which may be employed in the setting of advanced disease:
1. orchiectomy which removes the primary source of androgen production;
2. LHRH (luteinizing hormone releasing hormone) analogue therapy which depletes pituitary luteinizing hormone (LH) and results in subsequent down-regulation of LHRH receptors;
3. estrogen therapy which reduces LH production by an inhibitory effect at the hypothalamic/pituitary axis;
4. antiandrogen therapy which competitively inhibits androgen uptake at the cellular level; and
5. combined androgen blockade (CAB) where both androgen production and androgen uptake are inhibited.

Ninety percent of total testosterone is produced by the testis and 10% by the adrenal glands. Reduction of testicular testosterone can be accomplished with bilateral orchiectomy or by using LHRH agonists such as leuprolide acetate (Lupron®) or goserelin acetate (Zoladex®). These approaches are therapeutically equivalent. While orchiectomy is a very safe procedure, patients are increasingly reluctant to undergo this simple surgery, given the medical alternatives available. The LHRH analogues are available in a variety of depot preparations so that treatment (involving a subcutaneous injection) can be given as infrequently as once every four months. Side effects of a reduction in serum testosterone include loss of libido, impotence,

---

Urological Oncology, edited by Daniel Nachtsheim. ©2005 Landes Bioscience.

2

hot flashes, mild anemia (hemoglobin ~12 gm/dl), osteoporosis, and reduction in muscle mass. The efficacy of LHRH agonists in suppressing testicular production can be assessed by measuring serum testosterone. Castrate levels are <20 ng/ml. Using this approach, PSA will decline to "normal" (<4 ng/ml) in virtually all patients, and will be undetectable (<0.1 ng/ml) in approximately 20%. Regardless of which method of androgen ablation is employed, the median duration of this effect is approximately 30-36 months, a number which remain unchanged over the past 60 years.

Intermittent use of LHRH agonists has been advocated by some. Potential benefits include improved quality-of-life and a delay in development of a hormone refractory clone of cells. A large, randomized trial comparing continuous versus intermittent therapy is currently underway.

Antiandrogens are compounds which bind to target cell androgen receptors and prevent uptake of exogenous or endogenous androgens. They may be used alone or in combination with primary androgen suppression. Antiandrogen monotherapy without suppression of testicular androgens has been shown to produce inferior symptom-free interval and overall survival when compared to primary androgen suppression alone or CAB. There are two types of antiandrogens: nonsteroidal [flutamide (Eulexin®), nilutamide (Nilandron®) and bicalutamide (Casodex®)], and steroidal [cyproterone acetate (Androcur®) and megestrol acetate (Megace®)]. Steroidal antiandrogens are primarily used as second-line hormonal interventions, while nonsteroidal antiandrogens are used as primary therapy for recurrent or metastatic disease, either alone or in combination with orchiectomy or an LHRH agonist. The three commercially available nonsteroidal antiandrogens differ primarily in toxicity and half-life (Table 2.1).

Steroidal antiandrogens block the interaction between androgens and their receptors in target tissues. In addition, they have progestational activity that is, they lower LH production. Cyproterone acetate 250 mg/day was as efficacious as DES 3 mg/day in one study. Megestrol acetate is used less often than cyproterone acetate as primary therapy for metastatic disease. As monotherapy, neither agent alone has been found to suppress adrenal androgen completely, and after a couple of months serum testosterone levels rise to normal. The reason for this escape is poorly understood.

Adrenal androgens may be suppressed by the nonclassical antiandrogens, aminoglutethimide or ketoconazole. When these agents are employed, physiologic mineralocorticoid replacement must be employed. Typically, prednisone at a dose of 5 mg twice a day is sufficient. These agents are most commonly employed in the setting of progressive disease after initial hormonal intervention. Ketoconazole can be given to patients with spinal cord compression as initial therapy because it drops androgen levels to castrate levels within 24 hours. Long-term use is difficult due to toxicity and cost of this drug.

The 5α-reductase inhibitor finasteride (Proscar®) has been studied in advanced prostate cancer because of its inhibition of dihydrotestosterone production. It is also being studied as a chemoprevention agent based on the observation that men with a congenital 5α-reductase deficiency never develop prostate cancer. In metastatic disease, finasteride is often combined with an antiandrogen. The role of this agent in the treatment of prostate cancer remains to be defined.

Table 2.1. Antiandrogens commonly employed in the treatment of prostate cancer

Agent	Dose	Common Toxicities
Nonsteroidal		
Flutamide	250 mg TID	diarrhea, breast tenderness, nausea and vomiting
Bicalutamide	50 mg/day	nausea, breast tenderness
Nilutamide	300 mg/day	nausea, alcohol intolerance, decreased dark adaptation
Steroidal		
Megestrol acetate	40 mg QID	weight gain, dyspnea, edema
Cyproterone acetate	250 mg/day	nausea, decreased libido
Non-Classical		
Ketoconazole	200 mg TID	nausea, pruritus, impotency, nail dystrophy
Aminoglutethimide	1,000-1,750 mg/day	nausea, lethargy, rash, edema
Finasteride	5 mg/day	fatigue, hair growth

Combined Androgen Blockade

It was proposed in the 1980s that testicular androgen deprivation combined with an antiandrogen would be superior to testicular androgen deprivation alone. This is referred to as combined androgen blockade (CAB) or maximal androgen blockade (MAB). This concept is based on the theory that an antiandrogen will block adrenal androgens at the cellular level, enhancing the primary androgen suppression provided by orchiectomy or a LHRH agonist. There have been a number of large, prospective, randomized clinical trials which have compared CAB to monotherapy (primary androgen deprivation). A number of these trials demonstrated benefit for CAB, although the largest trial including 1,300 men randomized to orchiectomy with or without flutamide failed to demonstrate a difference in either progression-free or overall survival. Two meta-analyses examining all available data came to conflicting conclusions about the value of CAB. The issue remains unsettled, particularly when the additional variable of intermittent therapy is considered.

Timing of Initiation of Therapy

The optimal time to begin either monotherapy or CAB has not been settled. Earlier therapy results in more toxicity and increased costs. Loss of libido and muscle mass, impotence, anemia, hot flashes, hair loss, acceleration of osteoporosis, and fatigue are all potential toxicities of therapy.

When combined with radiation therapy, androgen deprivation appears to improve survival in patients with T3 or T4 lesions, as shown in the two studies in Table 2.2.

In patients with metastatic disease, early androgen deprivation is associated with improved symptom control and survival (Table 2.3). The most convincing data come from a study performed by the Medical Research Council. This study randomized 934 patients with locally advanced prostate cancer or asymptomatic metastasis to either immediate treatment (orchiectomy or LHRH agonist) or the same treatment initiated at the time of symptomatic progression. Results showed a more

Table 2.2. Adjuvant androgen deprivation and radiation therapy

Study	No. of Patients	Treatment	Survival	Comments
EORTC[1]	41% - 91% stage III	External beam XRT ± goserelin x 3 yrs	79% vs. 62% (p=0.001)	Androgen ablation at time of progression in XRT only group
RTOG[2]	73% stage III 26% stage IV	External beam XRT ± goserelin indefinitely	66% vs. 55% (p=0.03) Gleason 8-10 only	In LHRH arm all patients had improved rate of local recurrence, freedom from distant metastasis, disease-free survival and PSA relapse

[1] NEJM 1997; 337:295-300.
[2] J Clin Oncol 1997; 15:1013-1021.

rapid local and distant disease progression in the deferred group and a two-fold increase in serious complications.

Antiandrogen Withdrawal

Withdrawal of antiandrogens at progression is associated with a decline in PSA and improvement in clinical symptoms in approximately 20% of patients. The median duration of response is three to five months and is seen four weeks from withdrawal of flutamide. The time to response is longer in patients receiving bicalutamide, probably because of its longer half-life. A multivariate analysis has shown that important predictors of response included initial treatment with CAB, baseline alkaline phosphatase level and duration of therapy.

The mechanism of the antiandrogen withdrawal response is not fully understood; however, it appears to be at least partially related to mutation of the androgen receptor. The mutant receptors are paradoxically stimulated by antiandrogens. Recognition of this response is important and should be tried before initiating other therapies.

Secondary Hormonal Therapies

Despite the high initial response to hormonal therapy, virtually all patients with metastatic disease progress. The median time to progression is 16-18 months. When progression occurs, primary androgen deprivation should be maintained. This approach is supported by a retrospective review of the Eastern Cooperative Oncology Group database showing a superior overall survival for patients maintained on androgen deprivation. As discussed above, antiandrogen therapy should be discontinued. A number of agents have been tried and have been found to be of limited benefit. Suppression of adrenal androgens with ketoconazole (200-400 mg TID) or aminoglutethimide (1,000-1,750 mg/day) are associated with response rates of approximately 20% and disease stabilization in an additional 30%. Ketoconazole, in particular, has a very rapid onset of action and may be the initial treatment of choice for patients with extreme bone pain or spinal cord compression.

Table 2.3. Summary of the Medical Research Council prostate cancer working party investigator's trial

Complication	Immediate Treatment (n=469)	Deferred Treatment (n=465)
Pain from metastasis	121	211
Need for transurethral resection of the prostate	65	141
Pathologic fracture	11	21
Spinal cord compression	9	23
Ureteral obstruction	33	55
Extraskeletal metastasis	37	55
Death from prostate cancer	203	257

Br J Urol 1997; 79:235.

Estrogens and progestins have modest response rates in hormone refractory disease. Estrogens suppress pituitary gonadotropins, resulting in decreased testicular secretion of testosterone. In addition, estrogens have direct cytotoxic effects on tumor cells. Diethylstilbestrol, once the mainstay of therapy, is often difficult to obtain. Stilphostrol, a parenteral preparation, may be used at a dose of 1-1.5 gm/day for seven days, followed by the same dose, weekly. Other preparations of estrogen have not been well studied in this setting, and it should not be assumed they are equivalent to DES.

Progestins act by a mechanism which is unclear. Medroxyprogesterone (500-1,200 mg/day) and megestrol acetate (160-640 mg/day) have similar response rates, approximately 15%. There appears to be no advantage for doses of megestrol acetate higher than 160 mg/day.

Other drugs such as tamoxifen, somatostatin analogues, calcitriol, and retinoids have all been studied in the setting of hormone refractory disease. Responses to the first two have been observed in approximately 20% of patients, while there have been no clear responses to either calcitriol or all-trans-retinoic acid as single agents. Other retinoids are currently being studied.

Chemotherapy of Hormone Refractory Disease

Historically, hormone refractory prostate cancer has been viewed as a condition minimally responsive to cytotoxic agents. This view is largely based on data generated before the mid-1990s using single agents or combinations of drugs. Doxorubicin, cyclophosphamide, vinblastine, cisplatin, fluorouracil, mitoxantrone, and estramustine have all been shown to have objective response rates <15%. As a single agent, only estramustine is approved by the Food and Drug Administration (FDA) for the treatment of hormone refractory disease. Not only do these agents suffer from low response rates, they are particularly poorly tolerated in this group of frail, often poor performance status patients. Patients with hormone refractory disease are often elderly and much of their marrow has been irradiated making hematologic toxicity more likely. They often have comorbid medical conditions, and bony metastasis, if present, are often painful. Until the discovery and widespread use of PSA, objective assessment of response has been problematic. Several studies, both retrospective and prospective have demonstrated that a decline in PSA of at least 50% from baseline correlates with an objective response and may correlate with overall survival. In determining the activity of a therapy, a 50% PSA decline can be a useful surrogate marker. Ultimate utility of a given treatment still requires prospective, randomized comparisons.

The combination of mitoxantrone, a synthetic anthracenedione similar in structure and activity to doxorubicin but less toxic, and prednisone was recently studied. Patients receiving the combination were compared to those receiving prednisone alone. The combination of mitoxantrone (12 mg/m^2 every three weeks) and prednisone (10 mg/day) resulted in a highly significant improvement in palliation of symptoms (predominately bone pain) which led to the approval of this combination by the FDA. While the amount and duration of symptom palliation was both statistically and clinically significant, there was no impact of this therapy on overall survival. The median survival of both groups was 10 months, identical to historical median survivals.

More recent approaches have focussed on combined antimicrotublar therapy. Estramustine, vincristine, vinblastine, etoposide, and vinorelbine all prevent microtubule assembly during mitosis. The taxanes, docetaxel, and paclitaxel stabilize microtubules and inhibit microtubule disassembly. In vitro, the microtubule spindle inhibitors and taxanes are highly synergistic. The taxanes also inactivate bcl-2, an antiapoptotic protein expressed in 66% of hormone refractory prostate tumors. A number of clinical studies have been conducted employing estramustine and additional agents, some of which are summarized below. Estramustine doses and schedules vary widely (Table 2.4).

All of these regimens have response rates in the 40-60% range, which exceed those of older agents or combinations. The biochemical (PSA) and objective response rates appear to correlate. While these are all phase II studies, there is a suggestion that the historical median survival of 6-10 months may have been improved upon. The optimal dose and schedule of estramustine is not known. Two prospective, randomized studies in symptomatic, hormone refractory patients have been initiated. One study will compare docetaxel and estramustine given once every three weeks to mitoxantrone and prednisone given every three weeks. The second study will compare docetaxel and prednisone, given either weekly or once every three weeks to mitoxantrone and prednisone. Both studies will look at response rates, quality of life, toxicity and overall survival. A recently reported randomized phase II study comparing mitoxantrone and prednisone to estramustine, prednisone and either weekly or every three week docetaxel. Response, time to progression and toxicity all favor the taxane containing arms.[14]

Non-estramustine-based regimens have also been studied. Based on available data, it is not possible to conclude whether the activity of these combinations is equivalent to regimens which contain estramustine. At this time, it appears that docetaxel is the single most active agent in the treatment of hormone refractory disease. Chemotherapy offers definite promise to patients with this disease, but any impact on survival remains to be demonstrated.

Table 2.4. Estramustine based chemotherapy of hormone refractory prostate cancer

	Drug Combined with Estramustine	No. Patients	Response Rate	MST
Dimopoulos (1997)	Oral Etoposide 50 mg/m² x 21 days	56	5CR, 10PR/33 patients 58% >50% PSA decline	13 mos
Pienta (1997)	Oral Etoposide 50 mg/m² x 21 days	62	8 PR/15 patients 39% >50% PSA decline	56 wks
Pienta (1997)	Taxol Etoposide		57% >50% PSA decline	NR
Hudes (1997)	Taxol 120 mg/m² over 96 hours q 21 days	34	1CR, 3PR/9 patients 53% >50% PSA decline	69 wks
Colleoni (1997)	Oral Etoposide 50 mg/m² days 1-14 Vinorelbine 20 mg/m² days 1, 8, 28, 35	25	2PR/3 patients 56% >50% PSA decline	NR
Petrylak (1999)	Docetaxel 40-70 mg/m² q 21 days	34	5PR/18 patients 63% >50% PSA decline	23 mos
Savarese (1999)	Docetaxel 70 mg/m² q 21 days	47	1CR, 4PR/9 patients 69% >50% PSA decline	NR
Kosty (2000)	Docetaxel 43 mg/m²/wk 3 of 4 wks	35	1PR/5 patients 58% >50% PSA reduction	NR

Management of Bone Metastasis

Bone metastasis occurs in 97% of patients with metastatic disease and accounts for a large proportion of the complications of prostate cancer. Analgesics and external beam radiotherapy are the mainstay of therapy; however, bone-seeking radiopharmaceuticals have also been studied. Strontium-89 and samarium-153 both reduce pain and delay time to progression in patients with bony metastasis. Toxicity of these agents is largely hematologic, and thrombocytopenia can limit repeated use of strontium-89. Current studies are investigating the role of these agents in earlier stage disease or combined with chemotherapeutic agents.

Bisphosphonates (pamidronate, clodronate, and zoledronate) are analogues of pyrophosphates, which are naturally occurring inhibitors of bone resorption. Preliminary results with zoledronate (Zoneia®) suggest fewer skeletal events, less pain and improvement in osteopenia/osteoporosis.[15]

Selected Readings

1. The Leuprolide Study Group. Leuprolide vs. Diethylstilbestrol for metastatic prostate cancer. N Engl J Med 1984; 311:1281-1286. The first study showing equivalence of LHRH agonists to a standard therapy.
2. Loprinzi CL, Michalak JC, Quella SK et al. Megesterol acetate for the prevention of hot flashes. N Engl J Med 1994; 331:347-352. A good discussion of the utility of Megesterol acetate in the treatment of hot flashes. Other useful therapies are also reviewed.

3. Eisenberger MA, Blumenstein BA, Crawford ED et al. Bilateral orchiectomy with or without flutamide for metastatic prostate cancer. N Engl J Med 1998; 339:1036-1042. A negative study for combined androgen blockade vs. monotherapy.

4. Crawford ED, Eisenberger MA, McLeod DG et al. A controlled trial of leuprolide with and without flutamide in prostatic carcinoma. N Engl J Med 1989; 321:419-424. A positive study supporting CAB vs. monotherapy.

5. Prostate Cancer Trialists' Collaborative Group. Maximum androgen blockade in advanced prostate cancer: An overview of 22 randomized trials with 3,283 deaths in 5,710 patients. Lancet 1995; 346:265-269. A meta-analysis of CAB vs. monotherapy.

6. The Medical Research Council Prostate Cancer Working Party Investigators Group. Immediate versus deferred treatment for advanced prostatic cancer: Initial results of the Medical Research Council Trial. Br J Urol 1997; 79:235-246. An excellent study demonstrating the benefit of early vs. deferred treatment of metastatic prostate cancer.

7. Small EJ, Srinivas S. The antiandrogen withdrawal syndrome: Experience in a large cohort of unselected advanced prostate cancer patients. Cancer 1995; 76:1428-1434. A discussion of the antiandrogen withdrawal syndrome and its clinical utility.

8. Taylor CD, Elson P, Trump DL. Importance of continued testicular suppression in hormone-refractory prostate cancer. J Clin Oncol 1993; 11:2167-2172. An excellent paper giving data supporting the need for continued androgen suppression in the face of progressive disease.

9. Kelly WK, Scher HI, Mazumdar M et al. Prostate-specific antigen as a measure of disease outcome in metastatic hormone-refractory prostate cancer. J Clin Oncol 1993; 11:607-615. The initial report of the utility of PSA as a surrogate measure of treatment efficacy.

10. Dawson NA. Treatment of progressive metastatic prostate cancer. Oncology 1993; 7:17-24. A good review of second-line hormonal therapies.

11. Tannock IF, Osoba D, Stockler M et al. Chemotherapy with mitoxantrone plus prednisone or prednisone alone of symptomatic hormone-resistant prostate cancer. A Canadian randomized trial with palliative endpoints. J Clin Oncol 1996; 14:1756-1764. The trial which led to the FDA approval of mitoxantrone/prednisone for the treatment of symptomatic hormone-refractory prostate cancer.

12. Petrylak DP, MacArthur RB, O'Connor J et al. Phase I trial of docetaxel with estramustine in androgen-independent prostate cancer. J Clin Oncol 1999; 17:958-967. An interesting study with the best reported median survival to date in this group of patients.

13. Logothetis CJ, Vogelzang NJ. Current treatment and investigative approaches to the management of hormone-refractory prostate cancer. In: Vogelzang NJ, Scardino PT, Shipley WU, Coffey DS, eds. Comprehensive Textbook of Genitourinary Oncology, 2nd ed. Philadelphia: Lippincott Williams and Wilkins, 2000:862-869. A brief, up-to-date review of cytotoxic and investigational therapies of hormone-refractory disease.

14. Oudards S, Beuzeboc P, Orathe CM et al. Preliminary results of a phase II randomized trial of docetaxel, estramustine and prednisone—time schedules—versus mitozantrone and prednisone in patients with hormone refractory cancer. Proc Am Soc Clin Oncol 2002; 706:Abstract.

15. Saad F, Gleason DM, Murray R et al. A randomized placebo-controlled trial of zoledronic acid in patients with hormone-refractory metastatic prostate cancer. J Natl Cancer Inst 2002; 94:1458-68.

CHAPTER 3

Radiotherapy for Localized Prostate Cancer

Huan B. Giap and Prabhakar Tripuraneni

Excluding skin cancer, carcinoma of the prostate is the most common malignancy in males (about 20%), accounting for about 10% of all new cancer per year. When the disease is localized, either surgery or radiotherapy offers a good chance for cure. The optimal treatment of prostate cancer depends on pretreatment PSA, Gleason score, clinical stage, patient's performance status, quality of life issues, and the patient's wish. Results for treatment outcomes should have long follow-up since late recurrence and distant metastases are frequent. The two primary treatment options for organ-confined disease are prostatectomy and radiotherapy. Hormonal manipulation is useful for the advanced stage. Combination of hormonal manipulation and local treatment (surgery or radiotherapy) is under clinical investigation for patients with bulky localized disease and/or high risk. Candidates for definitive radiation therapy must have a confirmed pathological diagnosis of cancer that is clinically confined to the prostate and/or surrounding tissues (stages I, II, and III). High-risk patients should have a bone scan and computed tomographic scan negative for metastases, but staging laparotomy and lymph node dissection are not required. Prophylactic irradiation of clinically or pathologically uninvolved pelvic lymph nodes does not appear to improve overall survival. In addition, patients considered poor medical candidates for radical prostatectomy can be treated with acceptably low complications if care is given to delivery technique. Patients with extra-prostatic disease are usually not candidates for prostatectomy, and are best suited for radiotherapy. One popular method used to predict extra-prostatic disease is Partin's Table, which relies on PSA, Gleason score and clinical stage. Roach et al developed an equation to assess the risk of extra-prostatic disease:

% Risk of extraprostatic disease = 2/3 PSA + 10 x (Gleason Score – 6)

Difficulties in comparing surgery and radiotherapy are due to:
- Lack of well-designed randomized clinical trials comparing the two modalities.
- Retrospective analyses are bias since patients in radiation series are generally older and have more advanced disease.
- Patients in surgical series are pathologically staged, while patients in radiation series are clinically staged. Clinical staging with DRE and imaging studies tends to underestimate the extra-prostatic disease and lymph node status.
- Older radiation series use old techniques that deliver lower dose and treat larger volume of normal tissues.

Urological Oncology, edited by Daniel Nachtsheim. ©2005 Landes Bioscience.

Prognostic Factors after Radiation Therapy

Major prognostic factors affecting outcomes after radiation therapy are pretreatment PSA, stage, and Gleason score. The most important predictor of outcome is pre-treatment PSA, which correlates to the risk of nodal and distant metastasis. A retrospective review of patients treated with XRT showed cause-specific survival rates to be significantly different at 10 years by T-stage: T1 (80%), T2 (65%), T3 (50%), and T4 (20%). Ten-year overall survival rate by T-stage: T1 (65%), T2 (45%), T3 (40%), and T4 (15%). The most common system used in evaluating the degree of histological differentiation is the Gleason grade, which is based on the degree of differentiation of primary and secondary patterns. Each pattern has a score from 1 to 5, where 1 is the most well-differentiated and 5 is the most poorly differentiated. The two scores are combined to give a total Gleason score. Patients with higher Gleason scores have a higher risk of distant and nodal metastasis.

Other worse prognostic factors includes presence of lymph node, nuclear DNA ploidy, p53 mutation, history of TURP.

Combination of Total Androgen Suppression and Radiation Therapy

Combination of total androgen suppression (TAS) and XRT has been evaluated in a prospective randomized trial for high-risk patients. RTOG 86-10 study randomized 455 patients with stage T2b, T2c, T3 to XRT and TAS versus XRT alone. XRT consisted of 45 Gy to the pelvis followed by a cone down boost with additional 20-25 Gy. TAS consists of goserelin (Zoladex) 3.6 mg given subcutaneously every month and flutamide (Eulexin) 250 mg by mouth TID for 2 months before and during the XRT. At the median follow-up time of 4.5 year, the combined modality has better disease-free survival (36% versus 15%) and local control (54% versus 29%), but overall survival is similar. The criticisms of this study are that not all patients have PSA, the PSA cutoff at 4.0 for definition of failure, and short follow-up time.

The role of combined TAS and radiotherapy in patients with localized prostate and good prognostic factors is currently being investigated by RTOG 9408 prospective randomized trials. Patients with clinical stage T1-T2a are randomized to XRT alone versus XRT plus TAS. TAS is given 2 months prior to and concurrent with XRT. Endpoints are overall survival, biochemical outcome, and effect on sexual functions. For patients with large prostate volume, the rationale to use neoadjuvant TAS is to reduce the prostate volumes; hence, there may be potentially less morbidity. However, this must be weighed against the possible side effects of TAS (hot flashes, decrease in libido, impotence, decrease in muscle mass, and anemia), which can significantly affect the quality of life.

Conventional Radiation Therapy

Until the availability of CT scan in 1972, the "conventional" or 2-D approach to treatment planning was the mainstay in radiation therapy. The target volume was estimated from planar radiographs, which are largely anatomical, or bony landmarks. Usually, one or two pairs of opposed rectangular fields are used to treat a target area. All the fields are typically co-planar; i.e., the central axes of all the fields are on the same plane (Fig. 3.1). The treated volume is large, and much normal tissue is unnecessarily irradiated. An assumption was made that the patient shape is cylindrical; hence,

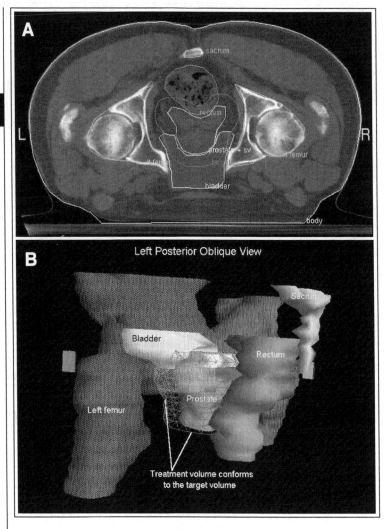

Fig. 3.1. 3-D reconstruction of pelvic organs and treatment volume which include prostate and seminal vesicle. (Nucletron Plato Treatment Planning System).

an outline of external contour of the patient was made in the midplane of the field using a lead wire or plaster cast. This outline was used for calculation of beam weighting and wedge. The isodose was generated for a single plane using this contour. The selection of plan is based on some rule of thumb and experience. The major limitations of this classical approach the are following:

1. The planar radiographs do not show precise extent of tumor or 3-D shape of the tumor;

2. The 3-D shapes of critical structures were not available;
3. The assumption that the patient shape is cylindrical is not valid especially for the head and neck and chest areas. The use of a single plane for plan evaluation is too simplistic;
4. The tissue inhomogeneities (lung, bone, air cavity) were not accounted for;
5. The dose algorithms did not take into account the 3-D body surface and changes in densities;
6. Plan evaluation was subjective; and
7. Since the fields were co-planar and rectangular, they are not tailored to the 3-D projection of the target volume.

The obstacles for advancing radiotherapy techniques are due to lack of 3-D imaging modality, linear accelerators with limited capabilities, non-availability of powerful computer systems, and lack of sophisticated dose calculation algorithms. Dosimetry calculation was done by hand until 1968, when computerized dosimetry was introduced. Axial images from CT for treatment planning were not used until 1976.

3-D Conformal Radiation Therapy

Significant advances have been made in the past two decades in the delivery of external beam radiation therapy, and the field of conventional radiotherapy has evolved into the three-dimensional conformal radiotherapy (3-D CRT) in the early 1990. The word "conformal" means that the high dose volume conforms or shapes to the target volume. The target volume is the tumor volume plus high-risk area (i.e., regional lymphatics) and a safety margin. The evolution of 3-D CRT is due to the improvements in the areas of beam delivery, 3-D imaging, patient immobilization techniques, computer hardware and software. These improvements have served several important purposes in the field of radiation therapy:

1. To allow the physician to define and treat the 3-D target volume more accurately and reliably;
2. To increase target dose, which hopefully translates into better tumor control;
3. To reduce the dose to normal tissue and volume of critical structures in the high-dose regions. This translates into reducing incidence and severity of radiation-related complications;
4. To automate the treatment delivery process so that the daily treatment is executed efficiently and reliably; and
5. To enhance our understanding of the radiobiology since the 3-D dose-volume relationship is known.

The process of 3-D CRT involves multiple steps that usually take about one to two weeks before the first treatment. On the first day of simulation, retrograde urethrogram (injection of radiopaque contrast into the penis) is done to define the inferior border of the prostate (apex). The patient is then set up in the prone position. A half-body plastic cast is fabricated as an immobilization device. Orthogonal radiographs are then taken to define the initial isocenter. The patient then undergoes a high-resolution CT scan of the pelvis for treatment planning. The digital CT images are transferred into a treatment planning computer. Next, the radiation oncologist outlines the normal critical structures and target volume on each CT slice. The initial target volume consists of prostate and seminal vesicle plus a margin of 7 to 15 mm. The margin accounts for organ motion and uncertainty in daily patient

3

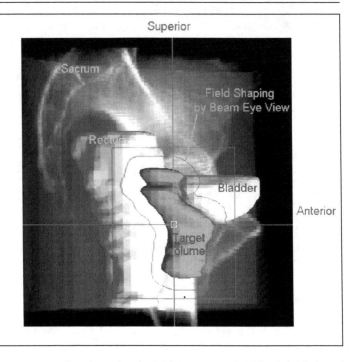

Fig. 3.2. Beam-Eye View is used to design the treatment portal for this right lateral field to treat the target volume which includes prostate and seminal vesicle, and to exclude most of the rectum and bladder out of the treatment portal. (Opti-rad treatment planning system, courtesy of Permedics, San Bernardino, CA).

setup. The pelvic lymphatics are typically not included in the target volume. Prophylactic irradiation of clinically or pathologically uninvolved pelvic lymph nodes does not appear to improve overall survival or cancer-specific survival, based on findings from a prospective randomized trial by RTOG. The next step in the process is treatment planning and optimization, in which medical dosimetrist/physicist works with the radiation oncologist to design the optimal configuration of the number beams and their corresponding weighting and shape. The 3-D reconstruction of the pelvic anatomy and target volume is done by "stacking" up the CT slices.

The beam angle and shape of radiation portal are designed by using the Beam Eye View (BEV) feature, which simulates the radiation source. Then, the angle is selected so that the entire target volume is included in the radiation portal while minimizing the volume of normal structure. An example of BEV is demonstrated in Figure 3.2.

The major advantage of 3-D conformal radiation therapy (3-D CRT) is to deliver a higher dose to the target volume and to reduce the volume of normal tissue irradiated. There are data from major centers suggesting that increase in dose to prostate (dose escalation) means increase in local control. Data from MSKCC showed the five-year biochemical relapse-free survival difference (92% versus 80%; p=0.03)

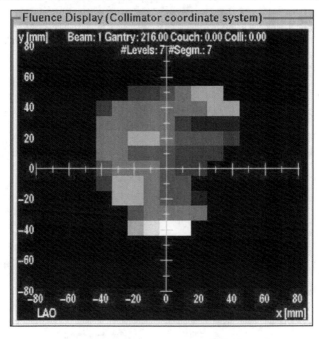

Fig. 3.3. Intensity-modulated radiation therapy (IMRT) utilizes 5-9 fields to sculpt the 3-D target volume. Each field has its own intensity map, which is painted by using dynamic motion of the multi-leaf collimator (MLC).

for patient receiving dose 75.6 Gy versus lower dose. Similarly, data from Fox Chase Cancer Center showed five-year biochemical relapse-free survival of 91% for dose >71.5 Gy versus 76% (p=0.01) for lower dose. Data from M.D. Anderson Cancer Center (MDACC) reported biochemical relapse-free survival of 95%, 85%, and 70% for dose >77 Gy, 67-77 Gy, and <67 Gy, respectively.

Intensity-Modulated Radiation Therapy (IMRT)

3-D CRT utilizes multiple beams at multiple angles to treat a target volume. Each beam has a single weighting. Intensity-modulated radiation therapy (IMRT) takes this one step further by adjusting the weighting (or intensity) of individual rays within each beam. By adding one additional level of control, IMRT produces superior dose distribution compared to 3-D CRT. The typical IMRT treatment utilizes 5-9 fields to sculpt the 3-D target volume. Each field has its own intensity map, which is painted by using dynamic motion of the multi-leaf collimator (MLC) (see Fig. 3.3). An irregularly shaped 3-D target volume can be treated by summing these fields. The result would be a more conformal dose distribution around the tumor volume, a sharper dose falloff at the boundary of the target volume, and less volume of normal tissue in the high dose region. IMRT evolves due to the availability of powerful computers, computer-controlled linear accelerators, advances in treatment planning software, dynamic multi-leaf collimators, and automation of treatment delivery.

Permanent Interstitial Brachytherapy

Permanent interstitial brachytherapy (or seed implant) has been used for patients with T1-T2 and favorable characteristics (low Gleason score and low PSA level). The initial experience with permanent seed implant was done with retropubic technique, which is more invasive and inferior in outcome. This has been replaced by a newer and more superior transperineal technique, which utilizes CT for planning/dosimetry and trans-rectal ultrasound (TRUS) to guide the implant. The transperineal implant is performed on an outpatient basis and does not require an open surgical procedure. About 100 Iodine-125 (I-125) and/or Palladium-103 (Pd-103) radioactive seeds with the total average activity of 50 mCi are typically used for the implant. MSKCC reported results for 109 patients with clinical stage T1-2 and Gleason score 4-7 who underwent CT-based transperineal I-125 implant alone from 1988-1995. With the median follow-up time of 4 years, the overall biochemical freedom from relapse (PSA <1.0) is 77% at 5 years. The biochemical local control is 100% for patients with pre-treatment PSA <4, 80% for pre-treatment PSA 4-10, <50% for patient with pre-treatment PSA >10. The University Community Hospital at Tampa, FL reported results for 124 clinically localized high-risk patients treated with Pd-103 seeds and external beam irradiation with a median follow up time of 3 years and biochemical control criterion of PSA <1.0. The high-risk patient is defined as pre-treatment PSA >10, Gleason score >6, or T2c or T3. The overall biochemical control is about 80%. One major advantage of seed implant is the high rate of preservation of sexual potency. Data from MSKCC reported potency preservation in 82% of the patients at 6 years, versus about 50% for patients treated with external beam irradiation alone. The data from Tampa FL, reported potency preservation of 77% at 3 years.

The most common acute morbidities of permanent seed implant are urinary retention and radiation urethritis. Urinary retention is more common in elderly patients with large prostate and pre-implant urinary obstructive symptoms. Radiation urethritis manifests as dysuria, daytime urinary frequency and nocturia. These acute symptoms are due to trauma from needle insertion and catheter use, and they usually start in the first week after implant, peak in the few weeks after, and generally resolve within several months. Temporary painful ejaculation is reported in about half of the patients, and it is probably due to the irritation of the terminal portion of the ejaculatory duct or the urethra. This symptom can be managed with NSAIDs, but it may require sexual abstinence for several months. Intermittent bloody semen (hematospermia) is often seen a few weeks after implant, and it is usually self-limited. Late hematospermia is also seen several years after implant, and it is probably due to radiation-induced capillary fragility. In patients without TURP history, significant urinary incontinence is uncommon, and the data for patients with TURP before or after implant are conflicting. Initial data at Northwest Tumor Institute (NWTI) reported 17% risk of post-implant urinary incontinence in patients with TURP history, and this may be due to more central loading of seeds. The data from Memorial Sloan-Kettering Cancer Center (MSKCC), which uses more peripheral loading, reported lower incidence (2 of 24 patients) (See Fig. 3.4).

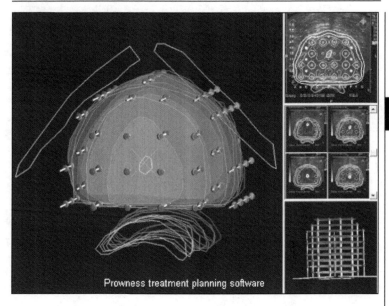

Fig. 3.4. Permanent seed implant (courtesy of Prowness Treatment Planning Software).

Temporary Interstitial Brachytherapy

Proponents for temporary interstitial implant cite the following advantages over the permanent seed implant:

1. Permanent seed implant delivers the radiation over a period of several weeks to months. During this period, the seeds may migrate or change position;
2. The prostate volume changes from pre-implant to a larger size right after the implant due to inflammation and then gradually decreases in size after that. Change in prostate size results in change in relative positions of seeds;
3. The seed placement is highly operator-dependent. Once the seeds are inserted, they cannot be adjusted;
4. The rapid fall-off of I-125 and Pd-103 due to the low energy, which may result in cold spots or under-dose;
5. The permanent seeds are short-lived, expensive and not re-usable; and
6. The permanent seed implant delivers at low-dose rate; hence, the biological equivalent dose is less.

There is some evidence that prostate carcinoma behaves like late reacting tissue; therefore, similar physical dose delivered at low dose rate is less effective.

Most temporary interstitial implants are done with Iridium-192 (Ir-192). Both low-dose rate (LDR) and high-dose rate (HDR) forms of Ir-192 have been used. The use of LDR Ir-192 has become less popular due to several reasons. First, LDR Ir-192 comes in seed ribbons with similar activity within a ribbon, and this does not

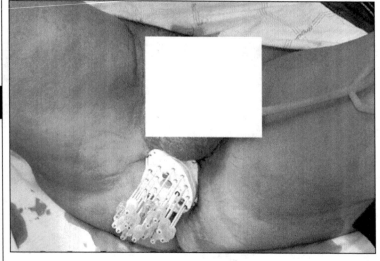

Fig. 3.5. Temporary interstitial brachytherapy.

allow spatial optimization. Second, LDR implants expose staffs to radiation exposure. Third, the biological equivalence dose to prostate tumor is higher for HDR than for LDR (see Fig. 3.5).

Temporary interstitial implant is currently done in combination with external beam irradiation (EBI). The optimal sequence has not yet been established; it can be either EBI first or the implant first. Most centers use EBI to deliver a dose of 36 to 50 Gy over 4 to 5 1/2 weeks to the limited pelvis or to prostate and seminal vesicles. The temporary interstitial implant is then given as the boost in a single or two implants, and one to three treatments are given per each implant. The optimal numbers of implants and the dose per fraction have not yet been established.

General eligibility criteria for HDR brachytherapy are:
1. Prostate volume <60 cc and good anatomy;
2. Patient is able to tolerate anesthesia;
3. No TURP in the last 3 months;
4. Clinical stage T2b-T3b and PSA <40 ng/ml; and
5. Clinical stage T1c-T2a, and Gleason score >6 or PSA >10.

Particle Beam Radiotherapy

Both proton and neutron particles have been clinically used in treatment of prostate cancer. The proton is a positively charged particle with the mass = 1,830 times of electron. The proton has finite range in matter. For a single beam in a homogeneous medium, the proton has superior dose distribution compared to photon or X-ray. As demonstrated in Figure 3.6, photons reach maximum dose deposition a few centimeters from the entrance and decrease from that on along the path of the beam. Along the track of the beam, the proton has less entrance dose, and the dose deposition increases with depth until it reaches the maximum at the end of its path

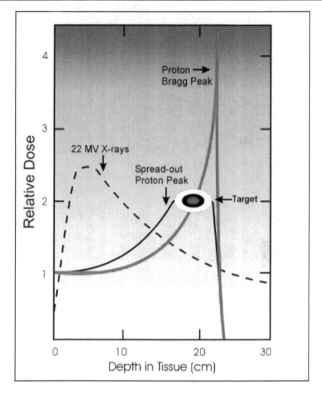

Fig. 3.6. Schematic demonstrating percent depth dose versus depth of tissue for proton versus photon.

(Bragg peak). The proton beam stops at the end of its Bragg peak while the photon beam continues to penetrate.

Proton beam also has less or equivalent side-to-side scatter (the sharper beam edge) to that of photon. The depth of the Bragg peak can be pre-determined by adjusted by the beam energy; for example, the 160 MeV proton beam has the range of 16 cm in water, and there is no dose beyond 16.5 cm. The width of the Bragg peak can also be adjusted by using a process called modulation. A 3-D bolus can also be used along a path of a single proton beam to conform the high dose to the 3-D target volume. In contrast, multiple photon beams must be used in 3-D conformal treatment. A typical treatment planning and dose distribution for proton beam are shown in Figure 3.7 (Images from the Optirad treatment planning system; courtesy of Permedics, Inc., San Bernadino, CA).

Proton has been used to treat prostate cancer since the late 1970 at Massachusetts General Hospital (MGH). However, limited technology restricted the use of proton beam in a large number of patients. The recent introduction of newer clinical facilities at Loma Linda University Medical Center and MGH have allowed

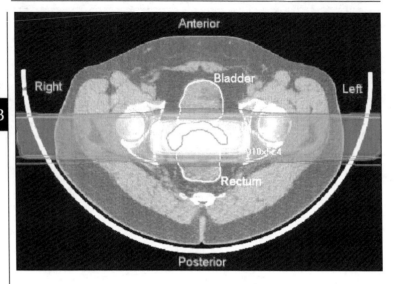

Fig. 3.7. A typical treatment plan and dose distribution for prostate using proton beam. (Opti-rad treatment planning system; courtesy of Permedics, San Bernadino, CA).

protons to be delivered to a larger number of patients. As of 2000, there are 19 proton treatment centers, and about 25,000 patients have received proton treatment worldwide. In contrast to the initial proton center in the physics department at Harvard University, which has a fixed beam and single energy of 160 MeV, these new hospital-based facilities have much better technologies such as higher and adjustable beam energy (250 MeV), uses of gantry instead of fixed beam to allow more complex beam angle, better patient immobilization and verification systems, and better treatment planning software. Slater et al reported Loma Linda University Medical Center results for 643 patients with localized prostate carcinoma treated with proton with and without photon for total dose of 74 to 75 CGE (Cobalt Gray Equivalent). The five-year biochemical disease-free survival (bDFS) rate was 89%. When stratifying the pre-treatment PSA, the bDFS is 100% for <4.0 ng/ml, 89% for 4.1 to 10.0 ng/ml, 72% for 10.1 to 20.0 ng/ml, and 53% for >20 ng/m. The post-treatment PSA nadir also carries prognostic significance, the bDFS is 91% for nadir <0.5 ng/ml versus 79% for nadir between 0.5 to 1.0 and 40% for nadir >1.0 ng/ml. The incidence of late rectal toxicity is 21% for RTOG grade 2 (rectal bleeding) and 0% grade 3. The incidence of late GU toxicity is 5.4% for grade 2 and 0.3% for grade 3. Shipley et al reported a phase III dose escalation study from MGH. Patients with locally advanced stage (T3-4) were randomized to dose of 64.8 Gy versus 75.6 Gy. Proton was used to deliver additional boost in the higher dose arm. There was no significant difference in survival between the two arms. There is improvement in local control with higher dose arm in patients with poorly differentiated histology. This study was done prior to availability of PSA. From reviews of published retrospective data, the outcomes for proton beam are equivalent to 3-D conformal radiotherapy and IMRT. Proton beam may be useful in dose-escalation studies.

PSA Follow-Up after Radiation Therapy

After radiation therapy, the PSA gradually decreases to nadir over 12-18 months. The half-life for PSA drop after XRT is about 2 months. It is controversial whether the time it takes to achieve PSA nadir and the half-life of PSA drop have any significant impact on outcome of XRT. The level of PSA nadir may play a factor in predicting outcome similar to the pre-treatment PSA. Several series have shown that the five-year biochemical relapse rate for patients with PSA nadir <1.0 µg/mop have any significant impact on the outcome of XRT. The level of PSA nadir may play a factor in predicting outcome similar to the pre-treatment PSA. Several series have shown that the five-year biochemical relapse rate for patients with PSA nadir <1.0 µg/ml is significantly lower than in patients with PSA nadir >1.0 µg/ml. Another controversy is the definition of biochemical relapse. Even though persistently elevated or rising PSA may be a prognostic factor for clinical disease recurrence, no definition has yet been shown to be an accurate surrogate for either clinical progression or survival. Reported case series have used a variety of definitions of "PSA failure" for overall survival is not known. As in the surgical series, many biochemical relapses (rising PSA alone) may not be clinically manifested in patients treated with radiation. In 1997, the American Society for Therapeutic Radiology and Oncology (ASTRO) consensus established that three consecutive PSA rises from nadir constitute biochemical relapse. The date of failure is the midpoint between the PSA nadir and the first of the three consecutive increases. There are several series showing the prognostic significance of the level of PSA nadir. Data from University of Texas at M.D. Anderson Cancer Center showed the five-year biochemical relapse rate is 17% for patients achieving PSA nadir (1.0 µg/ml versus 70% for patients with PSA nadir >1 µg/ml).

Acute Side Effects of Radiation Therapy

The Radiation therapy is usually well-tolerated. Definitive external beam radiation therapy can result in acute cystitis, urethritis, proctitis, and occasionally enteritis. These are generally reversible but may be chronic and rarely require surgical intervention. About two-thirds of patients will develop grade 2 or higher rectal and GU symptoms requiring medications. GU morbidity includes urinary frequency, urgency, hesitancy, dysuria, and nocturia. The symptoms usually start about the third week of XRT and resolve within a few weeks after completion of treatment. The GU symptoms can be managed by alpha-blockers (Flomax™, Cardura™, or Hytrin™), which significantly relieve the symptoms in two-thirds of patients, moderately improve the symptoms in additional one-fifth patients, and have no effect in 10% of patients. The dysuria can be controlled with Phanazopyridine HCl (Pyridium) and/or non-steroidal anti-inflammatory drugs (NSAIDs). The possibility of urinary tract infection should be ruled out since this is managed with appropriate antibiotics. Rectal morbidity includes rectal discomfort, tenesmus, and diarrhea. The symptoms of bowel urgency may be managed initially with diet manipulation and Metamucil. The bowel frequency and diarrhea can be managed with Imodium AD or Lomotil. Rectal irritation due to radiation proctitis can be managed similar to internal and external hemorrhoid by sitz baths, topical anesthetic and cortisone suppositories. Patients on anticoagulants (Coumadin, aspirin, etc.) may develop rectal bleeding, and they should have their blood work done and adjustment on the medication dosage adjusstments as indicated.

Late Complications of Radiation Therapy

The incidence of late complication is much lower, and these usually develop after 6 months up to 3-4 years after treatment. The two RTOG trials involving about 1,000 patients reported the 3.3% incidence of chronic GI sequelae (chronic diarrhea, rectal bleeding, rectal stricture) requiring hospitalization and minor intervention, 0.6% of the patient developing bowel obstruction or perforation. Rectal bleeding due to chronic proctitis is managed depending on severity of the symptom. Patients with intermittent mild to moderate rectal bleeding should be managed with mesalamine. Laser coagulation or cauterization could be considered for patient with persistent heavy bleeding.

The incidence of chronic GU sequelae (hematuria, urethral stricture, bladder contracture, or cystitis) requiring hospitalization is 7.3%, and 0.5% of the patients require major surgical intervention. More than half of the chronic urinary sequelae is urethral stricture, which occurs mostly in patients with history of TURP. Chronic urethritis, which usually manifests as urinary frequency and urgency can be treated with an alpha-blocker. The possibility of UTI or chronic prostatitis should be ruled out since both can be treated with appropriate antibiotics. UTI is usually associated with dysuria and can be detected by urinalysis and culture. The diagnosis of prostatitis can be made by DRE which typically reveals a tender and indurated prostate. The symptoms of frequency and urgency can also be due to radiation cystitis, which often associates with hematuria. If this is not responsive to conservative treatment, it can be treated with cauterization followed by silver nitrate or diluted formalin. The higher risk of complication is associated with higher dose (>70 Gy) and larger treatment volume.

Even though erectile potency is preserved in about 70-80% of patients at 12-15 months after XRT, this decreases with time. Only about 50% maintain erectile potency at 7 years. The etiology of erectile dysfunction after XRT was initially thought to be due to radiation damage to the nerve bundles; however, it is probably related to the damage to the vascular system, i.e., XRT-induced impotency is arteriogenic rather than cavernosal or neurogenic dysfunction. These patients can be managed initially with sildenafil (Viagra™) with 75% response rate, and patients with pre-XRT normal erectile function have better response than those with pre-XRT declining erectile function.

A population-based survey of Medicare recipients who had received surgery and radiation therapy as primary treatment of prostate cancer has been reported, showing substantial differences in post-treatment morbidity profiles between surgery and radiation. Although the men who had undergone radiation were older at the time of initial therapy, they were less likely to report the need for pads or clamps to control urinary wetness (7% versus more than 30%). A larger proportion of patients treated with radiation before surgery reported the ability to have an erection sufficient for intercourse in the month prior to the survey (men <70 years of age, 33% who received radiation versus 11% who underwent surgery alone; men (70 years of age, 27% who received radiation versus 12% who underwent surgery alone). However, men receiving radiation were more likely to report problems with bowel function, especially frequent bowel movements (10% versus 3%). Similar to the surgical patient survey, about 24% of radiation patients reported additional subsequent treatment of known or suspected cancer persistence or recurrence within 3 years of primary therapy.

Acknowledgments

The authors would like to acknowledge Robert Kirby from Permedics, Inc., for providing images from their Optirad treatment planning system, and Darrin Pella and Guolong Chu, Ph.D., for their assistance in providing the images from the Nucletron Plato treatment planning system.

Suggested Readings

1. Perez CA, Brady LW. "Principle and Practice of Radiation Oncology," 3rd ed. Philadelphia: Lippincott-Raven, 1998.

2. National Cancer Institute CancerNet PDQ® Cancer Information Summaries. Jan 2000. Website (http://cancernet.nci.nih.gov/pdq/) or phone 800-4-CANCER.

3. Vogelzang NJ, Scardino PT , Shipley WU et al. "Comprehensive Textbook of Genitourinary Oncology," Baltimore: Williams & Wilkins, 1996.

4. Webb S. The physics of three-dimensional radiation therapy. Philadelphia: Institute of Physics Publishing, 1993.

5. Cox JD, ed. "Moss' Radiation Oncology: Rationale, Technique, Results," 7th ed. St. Louis: Mosby, 1994.

6. Levitt SH, Khan FM, Potish RA, eds. "Levitt and Tapley's Technological Basis of Radiation Therapy: Practical Clinical Applications," 2nd ed. Philadelphia: Lea & Febiger, 1992.

7. Coia LR, Moylan DJ. "Introduction to Clinical Radiation Oncology," 2nd ed. Madison: Medical Physics Publishing, 1994.

References

1. Gleason DF, Mellinger GT. Prediction of prognosis for prostatic adenocarcinoma by combined histological grading and clinical staging. J Urol 1974; 111:58-64.

2. Pilepich MV, Caplan R, Byhardt RW et al. Phase III trial of androgen suppression using goserelin in unfavorable-prognosis carcinoma of the prostate treated with definitive radiotherapy: report of Radiation Therapy Oncology Group protocol 85-31. J Clin Oncol 1997; 15(3):1013-1021.

3. Webb S. The physics of three-dimensional radiation therapy. Philadelphia: Institute of Physics Publishing, 1993.

4. Asbell SO, Martz KL, Shin KH et al. Impact of surgical staging in evaluating the radiotherapeutic outcome in RTOG #77-06, a phase III study for T1BN0M0 (A2) and T2N0M0 (B) prostate carcinoma. Int J Radiat Oncol Biol Phys 1998; 40(4):769-782.

5. Wallner KE, Roy J, Harrison L. Tumor control and morbidity following transperineal I-125 implantation for stage T1-2 prostate carcinoma. J Clin Oncol 1996; 4:449-453.

6. Dattoli M, Wallner K, Sorace R. Pd-103 brachytherapy and external beam irradiation for clinically-localized high risk prostatic carcinoma. Int J Radiat Oncol Biol Phys 1996; 35:875-879.

7. Blasko JC, Wallner K. Brachytherapy for early prostate cancer. 41st ASTRO refresher course. Nov 1999. San Antonio, TX.

8.   Slater DS, Yonemoto LT, Rossi CJ et al. Conformal proton therapy for prostate carcinoma. Int J Radiat Oncol Biol Phys 1998; 42:299-304.

9.   Shipley W, Verhey L, Munzenrider J et al. Advanced prostate cancer: Results of a randomized comparative trials of high dose irradiation boosting with conformal protons compared with conventional dose irradiation using photons alone. Int J Radiat Oncol Biol Phys 1995; 32:3-12.

10.   American Society for Therapeutic Radiology and Oncology Consensus Panel. Consensus statement: guidelines for PSA following radiation therapy. Int J Radiat Oncol Biol Phys 1997; 37(5):1035-1041.

11.   Kuban DA, El-Mahdi AM, Schellhammer PF. Prostate-specific antigen for pretreatment prediction and posttreatment evaluation of outcome after definitive irradiation for prostate cancer. Int J Radiat Oncol Biol Phys 1995; 32(2): 307-316.

12.   Kavadi VS, Zagars GK, Pollack A et al. PSA after radiation therapy for clinically localized prostate cancer: Prognostic implication. Int J Radiat Oncol Biol Phys 1994; 30:279-287.

13.   Lawton CA, Wong M, Pilepich MW et al. Long-term sequelae following external beam irradiation for adenocarcinoma of the prostate: analysis of RTOG studies 7506 and 7706. Int J Radiat Oncol Biol Phys 1991; 21:935-936.

14.   Goldstein I, Feldman MI, Deckers PJ et al. Radiation associated impotence. J Am Med Assoc 1984; 215:903.

15.   Zelefsky MJ, Eid JF. Elucidating the etiology of erectile dysfunction after definitive therapy for prostatic cancer. Int J Radiat Oncol Biol Phys 1998; 40:129-133.

16.   Fowler FJ, Barry MJ, Lu-Yao G et al. Patient-reported complications and follow-up treatment after radical prostatectomy—the National Medicare experience: 1988-1990 (updated June 1993). Urology 1993; 42(6):622-629.

17.   Fowler FJ, Barry MJ, Lu-Yao G et al. Outcomes of external-beam radiation therapy for prostate cancer: a study of Medicare beneficiaries in three Surveillance, Epidemiology, and End Results areas. J Clin Oncol 1996; 14(8):2258-2265.

CHAPTER 4

Kidney Tumors

Michael Tran

Benign

With widespread use of CT (computed tomography) scans, benign tumors of the kidney are being found with increasing frequency. However, the most common benign mass found incidentally in the kidney are simple cysts, which are usually of little clinical significance. Common benign tumors include adenoma, oncocytoma, and angiomyolipoma.

Adenoma

The most common neoplasm of renal tubular epithelium are papillary adenoma. These lesions are usually discovered during autopsy with a frequency between 7 and 22%. They are located in the renal cortex and are typically <1 cm in size. Histochemical studies suggest they originate from distal tubular epithelium. Microscopic morphology of these lesions resembles low-grade papillary renal cell carcinoma and no reliable cytologic criteria can distinguish between the two. With the difficulty in differentiating these lesions histologically, it is even more problematic to make the diagnosis of these lesions clinically. Previously, it was thought that lesions smaller than 3 cm had a low propensity for metastasis. However, several reports have shown that size does not accurately predict metastatic potential. Thus, it would be prudent to treat solid lesions of renal parenchyma as malignant lesions pending histopathologic analysis.

Oncocytoma

Renal oncocytomas comprise approximately 5% of renal tubular neoplasms. Grossly, these lesions appear tan or light brown in color, round in shape with a well-defined fibrous capsule. A central stellate fibrous band projecting peripherally may also be present, usually in larger tumors. Histologically, they consist of large polygonal cells with abundant eosinophilic cytoplasm filled with mitochondria, referred to as oncocytes. Though the growth pattern is usually solid, about 5% may contain cysts and tubules. The cell of origin appears to be the intercalated cells of the collecting duct. They can become fairly large, with a median size of 6 cm in collected series. Bilateral tumors have been described in about 6% of cases, both synchronous and asynchronous. Oncocytomas are also found in other organs, including the adrenal, salivary, thyroid, and parathyroid glands.

These tumors are more common in males and are usually found in the seventh decade. Despite their size, they are usually asymptomatic and discovered incidentally. On CT, magnetic resonance imaging (MRI), or ultrasonography, the central stellate scar may be noticed, though this is not a pathognomonic finding. Because

Urological Oncology, edited by Daniel Nachtsheim. ©2005 Landes Bioscience.

true oncocytomas are benign, it has been suggested that partial nephrectomy would be appropriate treatment. However, these are dependent on an accurate preoperative diagnosis. Aspiration cytology is unable to exclude carcinoma in 20-30% of cases. In addition, carcinomas may contain foci of oncocytes and oncocytomas may have areas of malignant degeneration. Therefore, the decision to perform nephron-sparing surgery for oncocytomas should follow the same consideration of such surgery for carcinomas.

Angiomyolipoma

4

Angiomyolipomas of the kidney are benign tumors which can occur sporadically or in patients with tuberous sclerosis. The sporadic tumors are usually unilateral, whereas those associated with tuberous sclerosis are often bilateral. Approximately 17-20% of cases of angiomyolipomas occur in association with tuberous sclerosis. On the other hand, between 40 and 80% of patients with tuberous sclerosis will develop angiomyolipomas. Tuberous sclerosis is a syndrome consisting of a classic triad of epilepsy, mental retardation, and adenoma sebaceum. These patients may also have hamartomas of the brain, retina, heart, bone, lung, and kidney. There also appears to be an increased incidence of renal cell carcinomas with an incidence as high as 2%.

These tumors are generally unencapsulated, can extend into the collecting system or perirenal fat, and appear yellow to gray. They are composed of mature fat cells, smooth muscle cells, and abnormal blood vessels. Though distant metastasis has not been described, malignant degeneration to sarcoma has been reported.

Presenting symptoms include local discomfort from the mass effect of the tumor, flank or abdominal pain from sudden hemorrhage and even hemorrhagic shock from significant retroperitoneal bleeding. Radiographic imaging is very useful in making the diagnosis. Angiomyolipomas appear echogenic on ultrasound because of the fat-nonfat interfaces. The fat content is also characteristic on CT scan with Hounsfield units of -70 to -30, and on MRI with high intensity of T1-weighted images. Smaller tumors can be managed conservatively and followed with imaging annually. Tumors >4 cm that are asymptomatic may also be followed with imaging semiannually or annually but 50% of these tumors will grow. Selective angioembolization (preferred) or nephron sparing surgery should be considered for symptomatic or larger tumors. Finally, should the diagnosis be unclear by imaging, the lesion should be treated like a solid renal parenchymal lesion.

Other Benign Tumors

Benign tumors can arise from many other cell types in the kidney. Fibromas can occur within the parenchyma, capsule, or even perinephric tissue. Though benign, they are difficult to distinguish from fibrosarcomas. Leiomyomas are found in approximately 5% of autopsy series. They are usually <2 cm, often multiple, and subcapsular in location. Lipomas are rare and usually occur in middle-aged women. When they arise in perinephric tissue, they can become very large and excision may require nephrectomy. An interesting but rare tumor is the renin-secreting juxtaglomerular cell tumor, which arises from the afferent arterioles of the juxtaglomerular apparatus. These patients will present with diastolic hypertension, hypokalemia, elevated plasma renin activity, and elevated aldosterone level. They occur more commonly in young women who present with headaches. These tumors are small, between 2 and 3 cm, and are a curable cause of hypertension by excision.

Malignant

Renal Cell Carcinoma

Incidence and Etiology

In the United States, approximately 30,000 cases of renal cell carcinoma occurred in 1998. Of these cases, 11,000 deaths were predicted. The incidence has increased 43% from 1973 to 1995. Renal cell carcinomas account for 3% of all adult malignancies, not including skin cancers. The majority are found in the sixth to seventh decade. These tumors occur more commonly in males, with a male:female ratio of 2:1. Obesity and smoking appear to be risk factors, with relative risks of 3.6 and 2.3, respectively. There have been no definitive associations with occupational carcinogens. The etiology in the majority of cases remains unclear. Hereditary forms have been described, including hereditary papillary renal cell carcinoma, familial renal cell carcinoma with or without association with von Hippel-Lindau (VHL) disease. VHL disease is an autosomal dominant disease comprised of retinal and central nervous system (usually cerebellar and spinal) hemangioblastomas. Patients with VHL disease are at >100 relative risk of developing renal cell carcinoma. Though most renal cell carcinomas are unilateral, those associated with VHL disease are multifocal and bilateral in up to 75% of cases. These patients may also have pheochromocytomas, pancreatic islet cell tumors, and cysts of the kidney, pancreas, and epididymis. Acquired renal cystic disease has been associated with renal cell carcinoma. Chronic renal failure, regardless of whether the patient is on dialysis or not, leads to the development of acquired renal cysts. The relative risk of developing renal cell carcinoma in these patients is 32 times that of the general population. To a lesser degree, renal cell carcinoma has also been associated with tuberous sclerosis.

Molecular Genetics

Abnormalities on the short arm of chromosome 3 have been described in both sporadic and familial forms of renal cell carcinoma. More specifically, the VHL gene, a tumor suppressor gene, has been mapped by linkage analysis and in situ hybridization to the 3p25 locus. The VHL protein appears to be involved in regulating transcriptional elongation, and its loss results in increased transcriptional elongation and tumor growth. In addition, it has been shown that the VHL protein is able to regulate vascular endothelial growth factor (VEGF) production at the post-transcriptional level. Analysis of tumors from patients with VHL suggests that these patients have lost the wild-type allele inherited from their non-affected parent. Thus, in patients with VHL, a mutated gene is inherited and a second genetic event results in loss of heterozygosity, leading to tumor formation. Mutations of the VHL gene have been found in up to 75% of cases of sporadic clear cell carcinoma. In these patients, two acquired genetic events have occurred. Hypomethylation, an epigenetic change, can also lead to inactivation of the VHL gene. This has been demonstrated in 10-20% of clear cell carcinomas.

Patients with hereditary papillary renal cell carcinoma do not have alterations in the 3p region. Therefore, it appears as though the molecular genetics of papillary renal cell carcinoma is different from that of clear cell carcinoma. No characteristic oncogene abnormality has been demonstrated with renal cell carcinoma, though several candidates have been described, including c-myc, epidermal growth factor receptor, and HER-2.

Pathology

In the past, renal cell carcinomas were referred to as hypernephromas or Grawitz' tumors, based on the erroneous thought that they arose from adrenal rests within the kidney. Ultrastructural and immunohistochemical studies have shown that the majority arise from the epithelial cells of the proximal tubule, though some subtypes can arise from the distal tubule or collecting duct.

Grossly, these tumors are round, have a pseudocapsule comprised of compressed fibrous tissue and parenchyma, appear gray-white to yellow, and can be quite heterogeneous with areas of necrosis, hemorrhage, calcification and cyst formation. They have a propensity to extend into the renal vein and inferior vena cava as a thrombus.

The American Joint Committee on Cancer proposed the following classification of renal cell carcinoma in 1997. Clear cell carcinoma comprises 70% of cases. These tumors have cells with clear cytoplasm arranged in sheets, acinar or alveolar pattern with delicate branching vasculature. Foci of eosinophilic cytoplasm are common and were previously referred to as "granular." There may be both cystic and solid components. Five percent may have sarcomatoid changes. As mentioned, they are associated with mutations of the VHL gene. Papillary renal carcinoma accounts for 10-15% of cases. These tumors have also been called chromophilic renal cell carcinoma or tubulopapillary carcinoma. They are characterized by papillary architecture with variable cytoplasmic staining. The papillary cores may harbor psammoma bodies, foamy macrophages, and edema fluid. They are associated with trisomy of 3q, 7, 12, 16, 17, 20, and loss of the Y chromosome. The third type is chromophobe renal carcinoma and accounts for 5% of cases. These are usually solid with pale or eosinophilic cytoplasm, which contain microvesicles. There may be a "halo" around the nucleus. A rare variety is the collecting duct carcinoma, which has irregular channels lined with hobnail appearing atypical epithelium, an inflamed desmoplastic stroma. Demonstration of collecting duct origin is usually difficult. Tumors without a predominant pattern fall into an unclassified category. Sarcomatoid changes may occur in any histologic type and portends a poorer prognosis.

Clinical Presentation

Though the classic triad consists of pain, hematuria, and a flank mass, few patients present with all three. In fact, those that do usually have fairly advanced disease. Hematuria, either microscopic or gross, appears to be the most common finding, in up to 60% of cases. Patients may have generalized symptoms such as weight loss or fever. On physical exam, an abdominal mass may be palpable, and a varicocele, usually on the left, may also be present.

Renal cell carcinoma is associated with a wide variety of paraneoplastic syndromes. Stauffer syndrome refers to patients with elevated alkaline phosphatase, prolonged prothrombin time, increased ·2-globulin, and decreased albumin without having evidence of liver metastases. Normalization of hepatic function after nephrectomy is an important prognostic factor, with 88% survival >1 year. Persistence of abnormal liver function is an adverse prognostic factor, with <25% surviving two years. Hypertension occurs in 20-40% of patients. Though many patients with renal cell carcinoma have hyperreninemia, the hormone is usually inactive in these cases. Anemia can also be found in 20-40% of cases and may be a result of hematuria, decreased production, or increased destruction by the spleen. Fever and weight loss have been described in 20-30% of patients. Hypercalcemia, though less commonly seen, is interesting in that it may be secondary to production of a parathyroid hormone related peptide, similar to that described in lung cancer. Erythrocytosis, also uncommon, may be due to tumor production of erythropoietin. Other uncommon findings include amyloidosis, neuromyopathy, and elevated glucagon, insulin, α-fetoprotein, β-human chorionic gonadotropin, and true PTH.

Radiographic Imaging and Diagnosis

With widespread use of ultrasound and CT, an increasing number of renal cell carcinomas are being found incidentally. Classically, intravenous pyelography (IVP) with tomography is the study of choice in evaluating these patients, particularly when they present with hematuria. Renal masses will appear as nephrographic defects, more likely ill-defined versus renal cysts, which tend to be well-defined. Suspicion of a renal mass on IVP warrants further investigation by ultrasound or CT. In fact, IVP may miss smaller lesions <3 cm, especially when the lesions are located on the anterior or posterior.

Sonography is often used to distinguish cystic from solid lesions. It is more accurate than CT without contrast and should be used in patients with compromised renal function and those with contrast allergies. In addition, sonography does not expose the patient to radiation. Bosniak proposed criteria to characterize cystic lesions which correlate with concern for malignancy. On one end are simple cysts, which are anechoic, thin-walled, well-defined, round or oval, and have enhanced transmission beyond the distal thickened wall. On the other end are cystic lesions with internal or mural solid components, thickened wall, impaired transmission of sound, multiple septations, and extensive calcification. Complex cystic lesions and masses which appear frankly solid with variable echogenicity should be further characterized by CT. Sonography should not be used to stage patients with renal cell carcinoma; however, it may be valuable in patients with venous extension of tumor by delineating the superior extent of the thrombus. It should be kept in mind that no modality can reliably distinguish between blood and tumor thrombus.

CT is an excellent modality to evaluate renal lesions. It has an accuracy of 98% in detecting simple cysts, which do not enhance and appear homogeneous with an imperceptible wall. Features of a cystic malignancy include components with an attenuation greater than water, enhancement with contrast, mural nodules, thickened wall and internal calcification. Unless there is suspicion of metastatic disease, lymphoma, hamartoma, or an infectious process, a solid renal lesion which enhances should be considered a primary renal tumor, of which renal cell carcinoma is most common. CT is useful in determining renal vein involvement, vena caval extension, metastases to lymph nodes or liver, and invasion of adjacent organs. It is less accurate at demonstrating perinephric invasion. Overall, it is the most valuable modality and most cost-effective means to detect and stage renal lesions.

The sensitivity of magnetic resonance imaging (MRI) in evaluating renal lesions was initially questioned, but technological advances have improved its detection of smaller lesions. Renal cell carcinomas have variable intensity with the surrounding parenchyma and enhance with administration of gadolinium. Its ability to provide staging information is comparable with that of CT. MRI is an excellent means of determining the degree of venous tumor extension. Currently, because of higher cost, longer scan time and claustrophobia, its use is limited to patients with contrast allergies.

Prior to CT, angiography was used to make the definitive diagnosis. Renal cell carcinomas are characterized by neovascularity, arteriovenous fistulas, pooling of contrast media and accentuation of capsular vessels. Today, renal arteriography is mainly used to define vascular anatomy prior to nephron sparing surgery in a solitary kidney, or in conjunction with angioinfarction.

Radiographic imaging usually provides sufficient information before proceeding with treatment. The accuracy of fine needle aspiration biopsy for renal lesions is 80%. Thus, it yields useful information when it confirms the diagnosis of renal cell carcinoma and should be interpreted with caution when the results are normal or suggest a benign tumor. Its role is mainly in assessing an indeterminate renal cyst or when it is unclear whether the lesion is renal in origin.

Staging and Prognosis

Stage is the most significant prognostic factor in renal cell carcinoma. In addition to abdominal imaging, chest x-ray, bone scan, serum liver function tests, and calcium should be obtained. Though some have felt that a normal alkaline phosphatase can obviate the need for a bone scan, a recent study has demonstrated that alkaline phosphatase is a poor predictor of bony metastases. The Robson classification was described 30 years ago. Stage I tumors are confined within the capsule. Stage II tumors invade the perinephric fat or adrenal but are confined within Gerota's fascia. Stage III tumors involve lymph nodes or extend into the renal vein or inferior vena cava. Stage IV tumors invade adjacent organs or have metastasized. This staging scheme is not as accurate as the TNM classification proposed by the American Joint Committee on Cancer. Tumor size correlates with survival, and though using a tumor size of 7 cm to separate T1 from T2 tumors is arbitrary, it appears as though patients with treated T1 tumors have survival comparable to that of the general population. Renal vein invasion alone may not affect survival if adequately removed and inferior vena cava involvement may affect survival minimally. Survival is significantly diminished in patients with extensive nodal involvement, invasion of adjacent organs, or metastatic disease. It should be kept in mind that on preoperative imaging, enlarged lymph nodes may be reactive in up to 50% of cases and these patients should not be denied the opportunity to have therapeutic surgery.

Patient-related factors have also been reported to affect prognosis. Symptoms at presentation, weight loss, poor performance status, elevated erythrocyte sedimentation rate, hypercalcemia, and elevated alkaline phosphatase are negative prognostic factors.

There is little prospective information on the natural history of renal cell carcinoma. However, retrospective and anecdotal experience suggests that tumors detected at an earlier stage will become symptomatic or metastasize within 2-5 years. Recent analysis of the National Cancer Institute's Surveillance, Epidemiology, and End Results (SEER) program provided trends in survival of patients with renal cell carcinoma by comparing cases between 1975-1985 to those between 1985-1995. It appears that five-year survival has improved for both localized and advanced disease in whites while survival trends in blacks have shown little change or have declined. The reasons for this are not entirely clear.

Treatment of Localized Disease

Currently, neither chemotherapy nor radiation has a role in the treatment of localized renal cell carcinoma. In fact, radiation used alone can lead to radiation nephritis and has also been implicated as a risk factor in the development of renal cell carcinoma. Whether adjuvant radiation following surgery can decrease local recurrence rates remains controversial. Surgery is the mainstay of treatment for these patients.

It is presumed that radical nephrectomy provides better results than simple nephrectomy and is the standard treatment unless nephron sparing surgery is elected. A radical nephrectomy entails removal of the kidney and perinephric fat within Gerota's fascia. This provides for a better surgical margin, which is of prognostic value. Additionally, up to 25% of tumors will have invasion of the perinephric fat. Traditionally, the ipsilateral adrenal gland was also removed; however, it has since been demonstrated that routine adrenalectomy is not required unless the kidney is extensively involved or the tumor is located in the upper pole. Regional lymphadenectomy is usually performed. Its role in improving the survival of patients with lymph node involvement is controversial. Lymphadenectomy may be of benefit in patients with limited or microscopic nodal involvement, especially since there is no effective alternative treatment. With extended nodal disease, lymphadenectomy is unlikely to be beneficial given the unpredictable lymphatic drainage of the kidney and the likelihood of synchronous or metachronous metastatic disease.

A variety of surgical approaches have been used in performing a radical nephrectomy, with the choice being affected by the patient's habitus, size, and location of the tumor. The technique chosen should allow early access to the renal artery and vein to prevent excessive blood loss. An anterior subcostal approach is commonly used. It begins at the anterior axillary line and extends just across the midline, staying 1-2 fingerbreadths below the costal margin. This incision is generally performed intraperitoneally, can be bilateral if needed, and combined with a median sternotomy for access to the right atrium. It affords excellent access to the renal vessels but predisposes the patient to postoperative ileus. Large upper pole tumors are often accessed through a thoracoabdominal incision. It begins at the 8th or 9th intercostal space and can be extended down the midline. The peritoneum is usually entered though an extraperitoneal approach can be performed. This approach has a higher risk of pulmonary morbidity. Flank incisions have also been utilized, either above the 11th or 12th rib. It is more difficult to have good exposure to the renal pedicle and is best used for smaller or lower pole tumors.

Extension of tumor into the renal vein or inferior vena cava occurs in 4-10% of cases. With complete excision of the tumor thrombus, venous extension does not appear to be an adverse prognostic factor. Thrombi involving just the renal vein can be milked back before dividing the renal vein. Extension into the vena cava usually requires proximal and distal control with cavotomy to remove the thrombus. Cephalad extension to the level of the intrahepatic veins may require occlusion of the superior mesenteric artery and the porta hepatis. When the thrombus reaches the right atrium, cardiopulmonary bypass with hypothermic circulatory arrest may be necessary. If the tumor thrombus invades the wall of the vena cava, excision of the involved segment may be necessary. These patients usually have a poorer prognosis.

Although radical nephrectomy has served as the standard in treatment of localized renal cell carcinoma, partial nephrectomy or nephron sparing surgery also has a role in certain situations. Patients with compromised renal function, anatomical or functional solitary kidney, and bilateral renal tumors are candidates for partial nephrectomy, since radical nephrectomy would leave these patients anephric. Renal cell carcinomas arising in patients with VHL disease tend to be multiple, bilateral, and can recur. Thus, these patients may also benefit from partial nephrectomy. Partial nephrectomy is usually performed in vivo, and ex vivo surgery with autotransplantation is now rarely required. The surgical approach in these cases is often via a flank incision, which allows the kidney to be more superficial in the surgical field, providing better exposure than an anterior subcostal incision. In situ hypothermia is utilized to minimize ischemic injury to the kidney. Experience with partial nephrectomy has widened its indications to include patients with smaller tumors and a normal contralateral kidney. Use of partial nephrectomy in the treatment of tumors <4 cm has resulted in five-year survival of 90%. The disadvantage has been local recurrence, which occurs in approximately 5% of patients. This is likely due to unresected microscopic multifocal components.

Renal cell carcinoma can become fairly large and invade adjacent structures before symptoms arise. With the exception of invasion of the ipsilateral adrenal gland, these tumors are T4 lesions and have a poor prognosis, with five-year survival of <5%. However, resection of these lesions can be fairly morbid but may offer palliation from symptoms. En bloc resection of the kidney along with involved colon, small bowel, liver, spleen, or pancreas can be performed in carefully selected patients.

Treatment of Metastatic Disease

Approximately 30% of patients with renal cell carcinoma have metastatic disease at the time of diagnosis. Unfortunately, because surgery remains the only significant modality of treatment, these patients do poorly, with five-year survival of 10% or less. The most common sites of metastases are lymph nodes, lung, bone, liver, and central nervous system.

Radiation therapy can be used for palliation of metastatic lesions. It has been used alone or in conjunction with surgery in palliating metastatic lesions to the bone and central nervous system. Chemotherapy has had little success in the treatment of metastatic renal cell carcinoma. Of the drugs tried between 1983 and 1993, an objective response was seen in only 6.8% of patients. It has been suggested that high levels of expression of the multidrug resistance (MDR) gene product, p-glycoprotein, may be responsible by actively transporting drugs out of the cell. There is current interest in looking for ways to influence the action of p-glycoprotein.

Because effective systemic therapy is lacking, radical nephrectomy along with resection of the metastatic lesions has been utilized. However, no randomized, prospective trials have been performed to compare it with radical nephrectomy alone, making it unclear if resection of the metastatic lesion is of benefit. The patients who would seem most likely to benefit from such an approach would be those with solitary metastases. Five-year survival rates as high as 35% have been reported in these patients. Those with metachronous metastases have a better prognosis than those with synchronous metastases. Nephrectomy also has a palliative role in patients with severe pain, severe hemorrhage not manageable with angioinfarction, symptoms from paraneoplastic syndromes, or symptoms from compression of adjacent organs.

Most of the efforts toward developing effective systemic therapy have been focused at immunotherapy. The observation that renal cell carcinoma has significant variable doubling time, can regress spontaneously, and have a fairly protracted metastatic course, suggested that the immune system may have a role in regulating tumor growth. Interferon, a lymphokine, was the first substance to be widely used. Used alone, it has shown an overall response of approximately 10% and a complete response rate of 1%. The use of IL-2, or T cell growth factor, has shown more promising results. IL-2 monotherapy, given via intravenous bolus, continuous intravenous infusion, or subcutaneously has resulted in an overall response of approximately 15% and a complete response of approximately 4%. IL-2 has also been used in combination with interferon, lymphokine activated killer cells, and tumor infiltrating lymphocytes. Overall response has ranged from 16-35%, with complete response between 4-9%. Retrospective analysis of existing studies has suggested a role for adjunctive nephrectomy. Patients who did not have nephrectomy prior to immunotherapy did not fare as well as those who did. However, because of significant variability in patient selection, dosages, dosing schedules, and route of administration, randomized trials are still needed to sort out the most effective regimen. The use of IL-2 is associated with significant toxicity, including pulmonary edema, hypotension, renal insufficiency, and multiorgan system failure. Careful patient selection and experience has reduced mortality from 4% to <1%. Trials are ongoing to determine whether regimens aimed at decreasing toxicity can maintain efficacy.

Other Malignant Tumors

Sarcomas comprise between 1-3% of primary malignant tumors of the kidney. Differentiation of these tumors from renal cell carcinoma before surgery is difficult. In fact, histologically they may resemble renal cell carcinoma with sarcomatoid features. It is felt that the majority arise from the renal capsule. Leiomyosarcoma is the most common primary sarcoma of the kidney, accounting for 60% of cases. They tend to be encapsulated, compress rather than invade the surrounding parenchyma, and metastasize early. It is usually difficult to distinguish them from renal cell carcinoma by CT or MRI preoperatively, unless there is suggestion of capsular origin. However, similar to renal cell carcinoma, there is no clear benefit form radiation or chemotherapy and surgical excision is the mainstay of therapy. With a high likelihood of local recurrence and distant metastasis, these patients usually have a poor prognosis. There are many other renal sarcomas described in the literature and the most effective treatment is radical nephrectomy, with the possibility of neoadjuvant or adjuvant radiation or chemotherapy. Liposarcomas account for 20% of renal sarcomas and can be confused with angiomyolipomas.

Malignant fibrous histiocytomas are the most common soft tissue sarcomas in the elderly but are rarely renal in origin. Hemangiopericytomas arise from pericytes, cells which envelop the capillaries. They are associated with hypoglycemia and hypertension. Fibrosarcomas, rhabdomyosarcomas, osteosarcomas, and carcinosarcomas have also been reported to originate from the kidney.

In autopsy series, lymphomas are found to involve the kidney in up to 35% of cases. Although primary renal lymphoma is rare, it is important to keep it in mind in the differential diagnosis of a renal mass, as surgical intervention is not indicated. On CT, lymphoma may appear as multiple parenchymal nodules, a mass contiguous with lymph nodes, a diffusely infiltrative mass, or a solitary renal lesion. Leukemia also commonly affects the kidney, though renal involvement is usually clinically silent. Leukemic infiltration occurs in up to 67% of cases, is characterized by diffuse bilateral cortical infiltration, and does not typically affect renal function.

Finally, solid tumors may also metastasize to the kidney. At autopsy, solid tumors give rise to renal metastases in up to 12% of cases. These lesions are rarely discovered prior to death. Virtually every solid tumor has been described to metastasize to the kidney, though the three most common are lung, breast, and gastrointestinal.

Selected Readings

1. Chow WH, Devesa SS, Warren JL et al. Rising incidence of renal cell cancer in the United States. JAMA 1999; 281:1628-1631.
2. Storkel S, Eble JN, Adlakha K et al. Classification of renal cell carcinoma. Cancer 1997; 80:987-989.
3. Guinan P, Sobin LH, Algaba F et al. TNM staging of renal cell carcinoma. Cancer 1997; 80:992-993.
4. Srigley JR, Hutter RV, Gelb AB et al. Current prognostic factors—Renal cell carcinoma. Cancer 1997; 80:994-996.
5. Papanicolaou N. Urinary tract imaging and intervention: Basic principles. In: Walsh PC, Retik AL, Vaughan ED, Wein AJ, eds. Campbell's Urology. Philadelphia: W.B. Saunders, 1998:170-260.
6. Davidson AJ, Hartman DS, Choyke PL et al. Radiologic assessment of renal masses: Implications for patient care. Radiology 1997; 202:297-305.
7. Vogelzang NJ, Stadler WM. Kidney cancer. Lancet 1998; 352:1691-1696.
8. Belldegrun A, deKernion JB. Renal tumors. In: Walsh PC, Retik AL, Vaughan ED, Wein AJ, eds. Campbell's Urology. Philadelphia: W.B. Saunders, 1998:2283-2326.
9. Jennings SB, Linehan WM. Renal, perirenal, and ureteral neoplasms. In: Gillenwater JY, Grayhack JT, Howards SS, Duckett JW, eds. Adult and Pediatric Urology. St. Louis: Mosby, 1996:643-694.
10. Novick AC, Streem SG. Surgery of the kidney. In: Walsh PC, Retik AL, Vaughan ED, Wein AJ, eds. Campbell's Urology. Philadelphia: W.B. Saunders, 1998:2973-3061.
11. Glassberg KI. Renal dysplasia and cystic disease of the kidney. In: Walsh PC, Retik AL, Vaughan ED, Wein AJ, eds. Campbell's Urology. Philadelphia: W.B. Saunders, 1998:1757-1813.
12. Figlin RA. Renal cell carcinoma: Management of advanced disease. J Urol 1999; 161:381-387.

CHAPTER 5

Transitional Cell Carcinoma of the Ureter and Renal Pelvis

Donald A. Elmajian

Introduction

The American Cancer Society estimates that in the year 2000, roughly 2,300 Americans will be newly diagnosed with transitional cell carcinoma (TCC) of the ureter and that about 500 patients will die from this disease. Ureteral transitional cell carcinoma occurs in twice as many men as women, peaking somewhere in the seventh decade. Exact statistics regarding the incidence of renal pelvic transitional cell carcinoma are difficult to obtain because renal cell carcinoma and renal pelvic transitional cell carcinoma are reported together by the American Cancer Society. Renal pelvic transitional cell carcinoma is relatively uncommon in the United States and more than likely is no more prevalent than twice the rate of ureteral transitional cell carcinoma. Most of the medical literature discusses ureteral and renal pelvic transitional cell carcinoma together because of their obvious relation, and therefore both are included together in this handbook. Collectively one refers to these entities as upper tract transitional cell carcinoma to distinguish them from bladder cancer and tumors of the urinary tract distal to the bladder, which are frequently referred to as involving the lower urinary tract. Primary renal tumors will be discussed in a separate chapter.

Natural History

Bilateral upper tract disease, either synchronous or metachronous, occurs in 2% to 5% of patients with upper tract transitional cell carcinoma. Approximately 40% of patients (range 25% to 70%) with ureteral or renal transitional cell carcinoma will develop lower tract transitional cell carcinoma in the future, underscoring the need for bladder surveillance following definitive treatment. This may be because the renal pelvis and ureter are exposed only transiently to urinary carcinogens whereas the reservoir function of the bladder allows longer dwell times. Conversely, 2% to 4% of patients with transitional cell carcinoma of the lower urinary tract will develop upper tract transitional cell carcinoma following definitive treatment of their lower tract pathology. These tumors typically occur within 5 years of treatment and provide the rationale for ongoing surveillance of the upper tracts after treatment of the bladder. Late recurrences do occur. With respect to the ureter, in one study 73% of primary tumors were located in the distal ureter, 24% in the mid ureter, while 3% were located in the proximal ureter. Similar to the bladder, mapping studies reveal a high incidence of multifocality in the involved renal unit. This may be explained by the concept coined "field change disease" which has been applied to the urinary tract and refers to the propensity for transitional cell epithelium to undergo malignant transformation over time and space.

Risk Factors

A direct link between cigarette smoking and transitional cell carcinoma is well established. It is hypothesized that tobacco-related carcinogens excreted in the urine exert their cancer promoting effect by direct contact with the urothelium in a dose-response fashion. In a population-based, case-control study of 998 individuals, cigarette smoking was associated with a 3.1 increase in relative risk of ureteral and renal pelvic tumors, with a greater than 45-year smoking duration associated with a 7.2-fold increase in risk. It was estimated that 70% and 40% of upper tract transitional cell carcinoma was attributable to smoking in men and women, respectively. There is an inverse risk association as the time from smoking cessation increases, but the risk over time does not reach that of non-smokers.

There are other risk factors for ureteral transitional cell carcinoma that have been well described in the literature. Phenacetin use and abuse induces a characteristic thickening of subepithelial capillary basement membranes within the renal parenchyma called capillarosclerosis. This pathognomonic finding for phenacetin abuse should alert one to the possibility of an associated upper tract transitional cell carcinoma since phenacetin abuse is an established risk factor for this entity.

Acrolein, a metabolite of the commonly used chemotherapeutic agent cyclophosphamide, increases the risk of hemorrhagic cystitis and urothelial malignancy. Most oncologists give Mesna together with cyclophosphamide administration. This agent binds acrolein, preventing this adverse long-term event from occurring.

In the Balkan countries of eastern Europe, an endemic degenerative interstitial nephropathy occurs called Balkan Nephropathy. Interestingly, this nephropathy is associated with a 100- to 200-fold increased risk of upper tract transitional cell carcinoma. These tumors are typically low-grade, multifocal, and may be bilateral. The successful conservative management of these tumors has provided the impetus to manage "traditional" upper tract transitional cell carcinoma in highly select patients in a similar conservative manner as described later in this chapter.

Natural History

There is a direct correlation of grade and stage with survival in patients with upper tract transitional cell carcinoma. The TNM classification system of upper tract transitional cell carcinoma is shown in Table 5.1. A retrospective study of 54 patients with clinically localized upper tract transitional cell carcinoma divided patients into low and high stage based on depth of invasion (Ta/T1 vs. T2/T3). The authors found that 83% of the time low-grade (I/II) tumors were also low stage tumors and that high-grade (III/IV) tumors were also high stage tumors. Median survival of patients with low-grade vs high-grade tumors was 66.8 months vs. 14.1 months. Median survival of patients with low stage vs high stage tumors was 91.1 months vs. 12.9 months. These differences were statistically significant. Twenty-five point nine percent of the patients developed local recurrence and 53.7% developed distant metastasis.

A general rule of thumb is that when one component of the urothelium is involved with transitional cell carcinoma, the remaining uninvolved urothelium is at

---

Table 5.1. TNM staging for renal pelvis and ureter

Primary Tumor (T)

TX	Primary tumor cannot be assessed
T0	No evidence of primary tumor
Ta	Papillary noninvasive carcinoma
Tis	Carcinoma in situ
T1	Tumor invades subepithelial connective tissue
T2	Tumor invades the muscularis
T3	(For renal pelvis only) Tumor invades beyond muscularis into peripelvic fat or the renal parenchyma
T3	(For ureter only) Tumor invades beyond muscularis into periureteric fat
T4	Tumor invades adjacent organs, or through the kidney into the perinephric fat

Regional Lymph Nodes (N)*

NX	Regional lymph nodes cannot be assessed
N0	No regional lymph node metastasis
N1	Metastasis in a single lymph node, 2 cm or less in greatest dimension
N2	Metastasis in a single lymph node, more than 2 cm but not more than 5 cm in greatest dimension; or multiple lymph nodes, none more than 5 cm in greatest dimension
N3	Metastasis in a lymph node more than 5 cm in greatest dimension

*Note: Laterality does not affect the N classification

Distant Metastasis (M)

MX	Distant metastasis cannot be assessed
M0	No distant metastasis
M1	Distant metastasis

The American Joint Committee on Cancer (AJCC) TNM staging system for the renal pelvis and ureter. In: Sorbin LH, ed. TNM Classification of Malignant Tumors. 5th ed. Philadelphia: Lippincott-Raven Publisher, 1997.

---

risk (field change disease). For the purposes of this discussion, one may divide the urinary tract into the upper tract (renal pelvis and ureters) and lower tract (bladder and urethra). As stated, in the face of upper tract transitional cell carcinoma the bladder is at high risk for recurrence (about 40%). The converse is not true. That is, there is a low risk (2% to 4%) of upper tract recurrence in the face of lower tract transitional cell carcinoma. This suggests that "seeding" occurs from cephalad to caudad along the urinary tract. Two sets of data exist to support this concept. The first is clinically based. When patients undergo removal of the kidney for upper tract transitional cell carcinoma, there is a very high incidence of tumor recurrence in the remaining ureteral stump (30% to 60%). "Seeding" of the lower ureter may have occurred. To obviate this risk, the classic teaching is for a patient with ureteral or renal pelvis transitional cell carcinoma to undergo removal of the entire ureter as well as the kidney, including a cuff of bladder around the ipsilateral ureteral orifice (nephroureterectomy). If the ureter is left in place, this stump must be regularly evaluated. When transitional cell carcinoma of the bladder is diagnosed in a patient who has undergone removal of a kidney, it is imperative that the urologist assess the patient for the presence of a ureteral stump, and rule out the possibility that the origin of the bladder tumor is from this entity.

The concept of seeding from the upper tracts is supported molecularly as well. "Downstream" spread of molecular genetic alterations associated with transitional cell carcinoma has been documented. While no one alteration is known to cause disease in and of itself, together these abnormalities likely have adverse prognostic significance in terms of predisposition and/or progression of disease. Active study in this regard is ongoing.

As with most malignancies, hematogenous spread of transitional cell carcinoma can also occur. Involved organ systems typically include lung, liver, and bone. Nodal involvement via lymphatic spread is also possible, particularly with higher grade lesions. Low-grade tumors have an approximately 1% chance of lymph node involvement. Lymphatic drainage closely follows the venous drainage of the ureter. Thus, both the retroperitoneal lymph nodes on the affected side, including the inter-aortocaval nodes, and the ipsilateral pelvic-iliac lymph nodes are at risk. While good literature does not exist to support or refute the efficacy of lymphadenectomy for upper tract transitional cell carcinoma, practitioners of this procedure cite studies that document the survival benefit of an extended pelvic lymphadenectomy for transitional cell carcinoma of the bladder. They argue that surgical lymphadenectomy likely has the same benefit in upper tract transitional cell carcinoma. Lymphadenectomy obviously has staging value, although at present this may not be of benefit to patients because of a paucity of effective adjuvant treatment(s).

Diagnosis

Many patients with upper tract transitional cell carcinoma will have microscopic or gross hematuria. In the face of gross hematuria, the clots obtained are frequently long and thin, like one would anticipate from a clot in a ureter. Hematuria, regardless of the etiology, is an indication for a urologic workup. This typically will involve an intravenous pyelogram (IVP) followed by cystoscopy. Retrograde pyelography may be necessary in a patient who cannot receive intravenous contrast or in whom the ureters or renal pelves are poorly visualized. One may opt for a renal ultrasound instead of an IVP. In this instance retrograde pyelography will be required. In an actively bleeding patient, cystoscopy can localize the side of pathology by visualizing bloody efflux from the ureteral orifice.

Patients with upper tract transitional cell carcinoma may experience dull flank pain from progressive distension of the urinary tract secondary to gradual obstruction from the growing tumor. Alternatively, patients with upper tract transitional cell carcinoma can experience renal colic from acute urinary obstruction, usually secondary to blood clots. Renal colic, with or without hematuria, is a common presenting feature of several urologic entities. With the advent of the widespread availability of rapid image computerized tomography, and the proven efficacy of this modality in evaluating nephrolithiasis, the imaging workup may stray from the classic evaluation described above. It is imperative that the clinician visualize the entire urinary tract in a patient with hematuria, even if the initial study is diagnostic of non-malignant pathology. In certain circumstances the clinician will be tempted to accept a contrast enhanced CT as evidence that the uroepithelium is normal. Controversy exists regarding the efficacy of CT in this instance. Classic teaching dictates that this modality is inadequate in this regard. One may consider obtaining a KUB immediately after the CT, taking advantage of the contrast medium to visualize the uroepithelium.

Ureteral transitional cell carcinoma appears as a filling defect within or along the ureteral lumen. Alternatively, it may be represented as a point of complete obstruction involving the affected ureter. It may be confused with a radiolucent kidney stone. In this instance non-contrast enhanced CT scan will demonstrate the kidney stone to be opaque. Retrograde pyelography will often demonstrate a filling defect within the ureter. In this instance there may be characteristic dilation of the ureter immediately distal to the tumor such that contrast in this area appears to resemble a wine goblet (Bergman's sign).

A ureteral filling defect should prompt one to obtain urinary cytology from the affected side. Similar to urinary cytology for transitional cell carcinoma of the bladder, upper tract urinary cytology for ureteral transitional cell carcinoma lacks sensitivity, particularly in the face of low-grade tumors. While sensitivity is markedly improved for higher grade lesions and carcinoma in situ, specificity is lacking. When possible, brush biopsies of the lesion should be obtained either under direct vision via the ureteroscope or under fluoroscopic guidance. In one study comparing brush biopsy to cytology, sensitivity was improved from 48% to 72% and yielded a specificity of 94%. Another study prospectively compared brushings and cytology to ureteroscopy (with or without biopsy) and found that ureteroscopy was more accurate than the combination of brushings and cytology in diagnosing upper tract transitional cell carcinoma (58% vs. 83%). While there was some initial concern with respect to the negative impact preoperative ureteroscopy might have on treatment outcome, this has been found not to be the case. A recent study of 96 matched patients, half of whom underwent diagnostic ureteroscopy, found no difference in clinically apparent adverse events or long-term disease specific survival as a result of diagnostic ureteroscopy.

The standard workup of a ureteral filling defect therefore should consist of ureteral cytology, brush biopsy if possible, and ureteroscopy with or without a biopsy. The typical ureteroscopic appearance of ureteral or renal pelvic transitional cell carcinoma when associated with radiographic imaging suggestive of the same, in the face of negative cytology, brushings or biopsy, should prompt one to proceed with an appropriate treatment plan to manage the tumor even though tissue confirmation is not available.

An appropriate metastatic survey should be undertaken when upper transitional cell carcinoma is suspected or present. This should include a chest X-ray and serum chemistry. A CT scan of the abdomen and pelvis can give one an indication of the extent of the local disease but cannot adequately stage this lesion or determine whether the tumor is multifocal or invasive in nature. CT can also evaluate the liver for the presence of metastasis. CT is inaccurate in assessing regional nodal involvement unless the lymph nodes are massively involved. A bone scan would be indicated in the face of an elevated alkaline phosphatase or in a patient in whom one was unenthusiastic about treatment.

As the above discussion alludes to, accurate clinical staging of upper tract transitional cell carcinoma is lacking. In this sense ureteral and renal pelvic transitional cell carcinoma is similar to transitional cell carcinoma of the bladder. Depth of invasion is critically important when predicting prognosis and is poorly assessed in the preoperative setting. Tumor understaging in bladder cancer occurs in 40% of patients while overstaging occurs in 20% of patients. Therefore, in transitional cell carcinoma of the bladder, random biopsies of the urothelium to detect carcinoma in situ are part of the standard preoperative assessment. This is not practical in the evaluation of ureteral transitional cell carcinoma and rarely possible in renal pelvic transitional cell carcinoma, which is unfortunate given the known propensity for multifocality. In one study comparing endoscopic versus surgical diagnosis, pre-treatment tumor grade correlated with pathologic grade 90% of the time. In contrast, while there was good correlation between the preoperative and postoperative stages in low stage tumors, only 67% correlation existed amongst the high stage tumors. In this study low-grade tumors were also low stage over 85% of the time. Similarly, in a recent study of 45 upper tract lesions assessed endoscopically, correlation with respect to grade occurred in 78% of the tumors, while 45% of the tumors were upstaged at definitive treatment. Thus, unfortunately, the staging limitations well known in bladder transitional cell carcinoma exist in upper tract transitional cell carcinoma as well.

Treatment

Nephroureterectomy

The classic management of renal pelvic and ureteral transitional cell carcinoma, regardless of stage, grade, or location, is nephroureterectomy with excision of a cuff of bladder. When a stump of ureter is left behind, or when the periureteral bladder epithelium is not removed, the incidence of recurrent disease ranges from 30% to 75%. Complete removal of the ipsilateral upper urinary tract avoids the uncertainty associated with occult multifocal tumor or the presence of carcinoma in situ. This procedure can be performed via one or two incisions depending on the surgeon's choice. The bladder should be opened and the intramural ureter removed intravesically as opposed to extravesically. The specimen should be removed intact rather than with division of the ureter in an effort to prevent tumor spillage into the field of resection.

As previously discussed, the efficacy of en-bloc retroperitoneal and ipsilateral pelvic lymphadenectomy is not well established. It is apparent that patients with extensive regional lymph node involvement do poorly regardless of treatment. Furthermore, while an exact percentage of nodal metastasis in the face of pathologically low stage disease is not known, one can safely assume this number to be low given the excellent survival rates of this patient population. Thus, lymphadenectomy theoretically may only have a role in patients with moderate stage or grade tumors because it is this group who may harbor occult micrometastatic disease amenable to surgical extirpation. Unfortunately, given the limitations of preoperative staging, one may not be able to accurately identify this patient group. In a practical sense then, regional lymphadenectomy may be appropriate in healthy individuals with larger tumors, multifocal tumors, or moderate or poorly differentiated tumors. On the other hand, given the absence of effective adjuvant treatment, and limitations with respect to staging, one may argue that the added morbidity associated with lymphadenectomy in most circumstances is justified because of its presumed survival advantage in patients with micrometastatic disease.

Segmental Ureteral Surgery

The concept of segmental resection for ureteral transitional cell carcinoma began to gain popularity in the mid-1970s when the successful management of patients with Balkan nephropathy-induced ureteral transitional cell carcinoma was reported. Subsequently multiple reports regarding the efficacy of distal ureterectomy with a cuff of bladder for low-grade distal ureteral tumors were reported. Combined, these studies report an ipsilateral upper tract tumor recurrence rate of approximately 5.4%. In one larger study the ipsilateral tumor recurrence rate was 22%, but virtually all those tumors were multifocal. Survival rates in these studies were comparable to control patients who underwent nephroureterectomy. To be fair, other studies have reported less favorable ipsilateral tumor recurrence rates of between 25% and 40%. Not all of the patients in these reports, however, had low-grade unifocal distal tumors, making it difficult to evaluate the efficacy of segmental resection.

The lack of accurate preoperative clinical staging assumes paramount importance in planning segmental distal ureterectomy. Segmental resection of higher grade tumors or of tumors in the mid- or proximal ureter is associated with higher recurrence and poorer cancer-specific survival. One should be encouraged to remove the entire ureter distal to the lesion rather than perform segmental resection with reestablishment of ureteral continuity because of the high rate of recurrence distal to the tumor. This can frequently be accomplished with a psoas hitch and ureteroneocystostomy. If one elects to remove a longer length of ureter, then an intestinal segment can be utilized for ureteral substitution. Currently, unifocal low grade, low stage distal ureteral tumors can be safely handled via segmental resection. For lesions that do not fit this category, segmental resection should be reserved for highly select cases.

Ureteroscopic Treatment

Technologic advancements in fiberoptics have allowed urologists to gain ever-improving access to both the lower and upper urinary tracts. In pushing treatment boundaries, the efficacy of endoscopic management of upper urinary tract tumors has been evaluated. The initial studies involved patients with compromised renal function, a solitary or dominant kidney, or patients with bilateral disease. The initial favorable reports have allowed urologists expert in endourologic techniques to expand the indications to patients with normal renal function and normal contralateral renal units. The initial energy source utilized was electrocautery. Subsequent experience suggests that the neodymium:YAG laser and the holmium:YAG laser are at least equally efficacious. These small caliber laser fibers allow for the use of smaller optical instrumentation and are, therefore, associated with lower long-term ureteral sequelae.

As one would anticipate, the best results are obtained when treating low stage, low-grade, small distal ureteral lesions. The overall recurrence rate, not stratifying by grade or stage, is approximately 35% for ureteroscopic management of ureteral and renal pelvic tumors. One must remember that these studies involved highly select patients who had tumors amenable to this approach. Caution must be emphasized in managing larger, higher grade lesions, particularly if located high in the urinary tract.

In one study involving 49 renal units initially managed ureteroscopically, 8 patients thought to be appropriate for ureteroscopic management underwent nephroureterectomy based on the severity of their disease. Of the remaining 41 renal units, 28 (68%) were rendered tumor-free after a mean of 1.57 treatments. Recurrent disease occurred in 8 of these 28 renal units (29%). Of the 28 renal units, 24 (86%) were tumor-free at last followup. The mean followup of the entire population was 32.6 months. The overall tumor-free rate was 59%. The authors emphasize the importance of upper tract imaging studies and surveillance ureteroscopy to diagnose recurrent tumor as early as possible.

Percutaneous Treatment

The percutaneous approach to renal pelvic and proximal ureteral tumors is attractive because it allows for the use of larger instrumentation that can greatly facilitate resection. Initially developed to manage renal pelvic stones, until recently concerns regarding tumor implantation and spillage, coupled with the risk of bleeding limited its usefulness in this setting. As experience increased, select individuals underwent management of their tumors via this approach. To date, only one case of tract seeding has been reported, and that was during a diagnostic percutaneous procedure, not a therapeutic one. In Europe, it is not uncommon for the tract to be irradiated after the procedure.

In one study of 17 patients, complete percutaneous resection of tumor was accomplished in 17 of 18 renal units. Of the 18 renal units, 15 (83.3%) and 14 (77.8%) had low stage and low grade disease. Locally recurrent disease developed in 6 of the 18 units (33%). Two-thirds of these 6 patients had high grade disease at initial resection. Nephroureterectomy was performed in 2 of these patients, while repeat endoscopic management was successful in 3 of the remaining 4 units. One patient with high grade, high stage disease died of progressive and/or recurrent disease.

In another study involving complete percutaneous tumor resection in 26 patients with renal pelvic transitional cell carcinoma, 6 patients (23%) developed recurrent disease in the treated renal pelvis. Nephroureterectomy was necessary in 4 of the 6, while repeat percutaneous treatment was able to salvage the remaining 2 renal units. Two patients died of transitional cell carcinoma. Of the 26 patients, 11 (42%) had recurrence elsewhere in the urinary tract (most commonly in the bladder). Taken together, these and other studies indicate that percutaneous management can be accomplished. At present the high rates of local recurrence in this favorable patient population suggest that this approach should be reserved for highly selected patients.

Chemotherapy

Instillational chemotherapy has been utilized as an adjunctive treatment following both ureteroscopic and percutaneous management. While BCG and mitomycin C have been used most commonly, other agents including thiotepa, 5-fluorouracil, and interferon alpha have been utilized. These agents can be instilled via a percutaneous nephrostomy tube following percutaneous treatment. Alternatively, placing a ureteral stent or ureteral catheter into the affected renal unit and then instilling the agent into the bladder or catheter can create reflux. While some studies have clearly shown topical treatment to be of benefit in preventing recurrence, other studies have been unable to reproduce similar findings. Selection bias, coupled with non-randomization, is the likely explanation for the conflicting results. Prospective randomized trials will ultimately be required to define the role of this treatment approach.

Laparoscopic Treatment

The surgical removal of the kidney for transitional cell carcinoma involves the removal of the kidney and ureter, along with a cuff of bladder as described above. This typically is done by creating an incision in the abdominal cavity through which the tissue is removed. With the advent of minimally invasive surgical techniques, the efficacy and safety of laparoscopic nephroureterectomy is being evaluated. Once again, this does appear to be a potential option for select patients.

The most complete study to date compared 25 patients who underwent laparoscopic nephroureterectomy with 17 matched patients who underwent an open approach during the same time period. While laparoscopic surgery took longer, it was associated with less blood loss, lower analgesic requirement, earlier time to oral intake and return to normal activity, and less time in the hospital. Importantly, two retroperitoneal and one bony pelvic recurrence occurred in the laparoscopic group, while none occurred in the group undergoing open surgical resection. While technically feasible, this technique requires further evaluation to firmly establish efficacy.

Summary

Transitional cell carcinoma of the upper urinary tract is uncommon. Cigarette smoking is the most significant risk factor for the development of this entity. Recurrence in the bladder or in the retained portion of the affected renal unit is very high and, therefore, careful surveillance posttreatment is emphasized. An appropriate evaluation in the face of hematuria, flank pain, or other symptoms suggestive of genitourinary pathology should involve the urologist and consist of imaging studies, cystoscopy, cytology, and ureteroscopy. The standard of care to manage these lesions involves open surgical resection. For select patients segmental resection and endoscopic or laparoscopic approaches may be appropriate, particularly if the tumor is small, and of low stage and grade. Preoperative staging is problematic. High stage tumors are frequently fatal if managed conservatively or left incompletely resected.

Selected Readings

1. Greenlee RT, Taylor M, Bolden S et al. Cancer Statistics, 2000. CA Cancer J Clin 2000; 50:7-33.
2. Babaian RJ, Johnson DE. Primary carcinoma of the ureter. J Urol 1980; 123(3):357-359.
3. Murphy DM, Zincke H, Furlow WL. Management of high grade transitional cell cancer of the upper urinary tract. J Urol 1981; 125(1):25-29.
4. Anderstrom C, Johansson SL, Pettersson S et al. Carcinoma of the ureter: A clinicopathologic study in 49 cases. J Urol 1989; 142(2 Pt 1):280-283.
5. Strong DW, Pearse HD. Recurrent urothelial tumors following surgery for transitional cell carcinoma of the upper urinary tract. Cancer 1976; 38(5):2173-2183.
6. Malkowicz SB, Skinner DG. Development of upper tract carcinoma after cystectomy for bladder carcinoma. Urology 1990; 36(1):20-22.
7. McLaughlin JK, Silverman DT, Hsing AW et al. Cigarette smoking and cancers of the pelvis and ureter. Cancer Res 1992; 52(2):254-257.

8.   Huben RP, Mounzer AM, Murphy GP. Tumor grade and stage as prognostic variables in upper tract urothelial tumors. Cancer 1988; 62(9):2016-2020.

9.   Dodd LG, Johnston WW, Robertson CN et al. Endoscopic brush cytology of the upper urinary tract. Evaluation of its efficacy and potential limitations in diagnosis. Acta Cytol 1997; 41(2):377-384.

10. Streem SB, Pontes JE, Novick AC et al. Ureteropyeloscopy in the evaluation of upper tract filling defects. J Urol 1986; 136(2):383-385.

11. Hendin BN, Streem SB, Levin HS et al. Impact of diagnostic ureteroscopy on long-term survival in patients with upper tract transitional cell carcinoma. J Urol 1999; 161(3):783-785.

12. Keeley FX, Kulp DA, Bibbo M et al. Diagnostic accuracy of ureteroscopic biopsy in upper tract transitional cell carcinoma. J Urol 1997; 157(1):33-37.

13. Guarnizo E, Pavlovich CP, Seiba M et al. Ureteroscopic biopsy of upper tract urothelial carcinoma: Improved diagnostic accuracy and histopathological considerations using a multi-biopsy approach. J Urol 2000; 163(1):52-55.

14. Zoretic S, Gonzales J. Primary carcinoma of ureters. Urology 1983; 21(4):354-356.

15. Witters S, Vereecken RL, Baert L et al. Primary neoplasm of the ureter: A review of twenty-eight cases. Eur Urol 1987; 13(4):256-258.

16. Murphy DM, Zincke H, Furlow WL. Primary grade I transitional cell carcinoma of the renal pelvis and ureter. J Urol 1980; 123(5):629-631.

17. Mufti GR, Gove JR, Badenoch DF et al. Transitional cell carcinoma of the renal pelvis and ureter. Br J Urol 1989; 63(2):135-140.

18. Ghazi MR, Morales PA, Al-Askari S. Primary carcinoma of the ureter. Report of 27 new cases. Urology 1979; 14(1):18-21.

19. Wallace DMA, Wallace DM, Whitfield HN et al. The late results of conservative surgery for upper tract urothelial carcinomas. Br J Urol 1981; 53:537-541.

20. Hatch TR, Hefty TR, Barry JM. Time-related recurrence rates inpatients with upper tract transitional cell carcinoma. J Urol 1988; 140:40-41.

21. Clark PE, Streem SB. Endourologic Management of Upper Tract Transitional Cell Carcinoma. AUA Update Series Lesson 16; Volume XVII:122-128.

22. Keeley FX Jr, Bibbo M, Bagley DH. Ureteroscopic treatment and surveillance of upper urinary tract transitional cell carcinoma. J Urol 1997; 157(5):1560-1565.

23. Clark PE, Streem SB, Geisinger MA. 13-year experience with percutaneous management of upper tract transitional cell carcinoma. J Urol 1999; 161(3):772-775; discussion, 775-776.

24. Patel A, Soonawalla P, Shepherd SF et al. Long-term outcome after percutaneous treatment of transitional cell carcinoma of the renal pelvis. J Urol 1996; 155:868-874.

25. Shalhav AL, Dunn MD, Portis AJ et al. Laparoscopic nephroureterectomy for upper tract transitional cell cancer: The Washington University experience. J Urol 2000; 163:1100-1104.

CHAPTER 6

Bladder Cancer

Jonathan E. Bernie and Joseph D. Schmidt

Introduction

Bladder cancer is the most common tumor of the urothelium. Most of these tumors are transitional cell carcinomas and are generally categorized as either low-grade and superficial or high-grade and invasive. Therapy for bladder cancer is evolving, but the most common modalities for treatment currently include—endoscopic tumor resection, intravesical chemotherapy, immunotherapy, cystectomy, and systemic chemotherapy.

Epidemiology

Approximately 54,500 (1999 estimate) new cases are diagnosed annually in the United States, with a male-to-female predominance of 3 to 1 and an incidence of 18 in every 100,000 in the population. In older men, bladder cancer accounts for 6% of all cancer cases and is the fourth most common malignancy (after prostate, lung, and colorectal). Bladder cancer in women is the eighth most common cancer. The incidence of bladder cancer is higher with advancing age and is highly variable throughout the world, with higher rates in Western Europe and North America and lower rates in Eastern Europe and some Asian countries. Five-year survival also differs between gender and race (white men, 84%; black men, 71%; white women, 76%; black women, 51%). This is primarily due to advanced stage at diagnosis, which is possibly related to delayed diagnosis, disparate tumor biology or access to health care. The risk of acquiring bladder cancer in one's lifetime is estimated to be 2.8% (white men), 0.9% (black men), 1% (white women), 0.6% (black women). Nearly 12,000 people die of the disease annually. The peak incidence of bladder cancer is the seventh decade of life. Eighty percent of patients are in the 50- to 79-year age range.

Risk Factors

Genetics

Numerous factors have been reported to be associated with bladder cancer. Although some genetic abnormalities have been associated with bladder cancer (p53, Rb, erbB-2, and loci on chromosome 9), these findings have not been uniform. Certain environmental exposures have been shown to correlate with bladder cancer.

Smoking

The incidence of bladder cancer is up to four-fold in cigarette smokers and many cases of bladder cancer are thought to be related to smoking. It has been suggested

Urological Oncology, edited by Daniel Nachtsheim. ©2005 Landes Bioscience.

Table 6.1. *Exposures/occupations at higher risk for bladder cancer*

Transitional Cell Carcinoma
Cigarette smoking
Occupations (rubber workers, textile and dye workers, painters)
Chemicals (benzidine, nitrosamines, 2-naphthylamine, 4-nitrobiphenyl,
    4-4-diaminobiphenyl)
Other (phenacetin, cyclophosphamide)

Squamous Cell Carcinoma
Schistosoma haematobium
Inflammation/infection (chronic indwelling catheters, bladder calculi,
    bladder diverticula)
Pelvic radiation

6

that reformed smokers do not reduce their risk to baseline until at least 20 years of abstinence.

Environmental Exposures

Aniline dyes (ortho-amino-phenols, e.g., 2-naphthylamine, 4-aminobiphenyl, 4-nitrobiphenyl, 4-4-diaminobiphenyl) and some aldehydes (acrolein) are also carcinogens which have been shown to correlate with an increased incidence of bladder cancer. Some occupations that are reported to be at higher risk for the disease include dry cleaners, painters, and leather workers. High-dose consumption of phenacetin (which has a chemical composition similar to aniline dyes) has also been correlated with urothelial cancers.

Chronic Inflammation

Chronic irritation and inflammation of the bladder is another risk factor for the development of bladder cancer. Patients with chronic indwelling catheters (spinal cord injured patients) are at higher risk for the development of squamous cell carcinoma of the bladder. Also, *Schistosoma haematobium* infection (more common in Egypt) is also associated with squamous cell bladder cancer. Patients treated with cyclophosphamide (Cytoxan®) are also at increased risk. It has been suggested that 2-mercaptoethanesulfonic acid (Mesna®) may provide some protection from cyclophosphamide-induced cancer.

Other

In the past it had been suggested that coffee drinking and artificial sweeteners were associated with bladder cancer. Further analysis of the data does not support these claims (see Table 6.1).

Natural History

Most bladder cancers newly diagnosed in the United States (70%) are well to moderately differentiated, papillary, and superficial. Up to 70% will recur after local resection, and about 15% of these patients will eventually develop muscle invasive or metastatic bladder cancer. Prognostic factors include histologic grade, stage, vascular/lymphatic invasion, presence of carcinoma in situ (CIS), and possibly smoking. Increased risk of recurrence is found in patients with multiple tumors, CIS, lamina propria invasion, and grade III disease. The remaining 30% of patients have high-grade and/or muscle invasive tumors at presentation. Of the patients with muscle invasive disease, about half will have distant metastases. Five-year survival for patients with regional lymph node metastases is approximately 20-50%. Almost all patients with distant metastatic disease will die within two years, with the most optimistic reported five-year survival as 6%.

Classification

Anatomy

The normal bladder epithelium consists of transitional cell lining, up to seven cell layers thick. Deep to the urothelium is the subepithelial connective tissue (lamina propria or submucosa) which contains irregularly arranged smooth muscle fibers. The muscularis (detrusor muscle) is adjacent to the lamina propria and is surrounded by perivesical fat.

Microscopic Description

The tumor histology of most bladder cancers is (90%) transitional cell carcinoma. Typically, these tumors will have an increased number of cell layers, increased nuclear-to-cytoplasmic ratio, increased mitoses, and prominent nucleoli. The tumor growth can be papillary, sessile, nodular, mixed or flat (carcinoma in situ, CIS). Approximately two-thirds of all bladder cancers will be of the papillary variety.

Carcinoma in situ is high-grade transitional cell carcinoma that is confined to the superficial urothelium. It occurs more commonly in men and can create symptoms that mimic prostatism and bladder irritability. Cystoscopically the mucosa can appear erythematous and is classically described as velvety in appearance. Patients with CIS have higher tumor recurrence rates and often have positive urine cytologies (up to 90%). CIS can occur in association in up to 75% of muscle invasive cancers.

Up to 20% of patients with bladder cancer have lymphatic or vascular metastases. Lymphatic spread usually occurs to the pelvic lymph nodes. Hematogenous metastases most commonly involve the following sites—liver, lung, bone, and adrenal.

Staging/Grading

The first staging classification was set forth by Jewett and Strong in 1946 and still forms much of the basis for the current TNM staging for bladder cancer. The basic principle is that depth of bladder wall invasion inversely correlates with survival. "T" describes the extent and depth of the primary tumor, "N" describes the presence, number, and size of lymph node involvement and "M" details the presence or absence of distant disease.

Clinical Staging

The assessment of the primary tumor includes histologic diagnosis/verification and bimanual examination under anesthesia. A thickened bladder wall, mobile mass or fixed mass found on bimanual examination strongly implies the presence of T3a or greater disease.

6

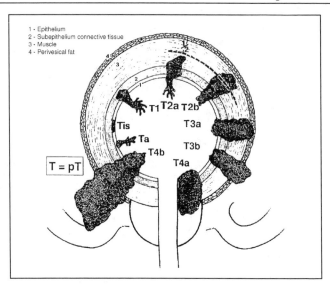

Fig. 6.1. Illustration of TNM/UICC staging of bladder tumors. T0, no evidence of primary tumor; Ta, non-invasive papillary carcinoma; Tis, carcinoma in situ: "flat tumor"; T1, tumor invades subepithelial connective tissue; T2, tumor invades muscle; T2a, tumor invades superficial muscle (inner half); T2b, tumor invades deep muscle (outer half); T3, tumor invades perivesical tissue; T3a, microscopically; T3b, macroscopically (extravesical mass); T4, tumor invades any of the following: prostate, uterus, vagina, pelvic wall, abdominal wall; T4a, tumor invades prostate or uterus or vagina; T4b, tumor invades pelvic wall or abdominal wall. Reprinted with permission from Hermanek P, Hutter RVP, Sobin LH et al, eds. UICC TNM Atlas, 4th ed. Berlin, Heidelberg, New York: Springer-Verlag, 1997©.

Pathologic Staging

Histopathologic examination is necessary. Cystectomy and nodal specimens are usually required to provide adequate staging. Nodal status is determined both by number and size of positive lymph nodes. The regional nodes of the true pelvis include hypogastric, obturator, perivesical, pelvic, sacral, presacral, and iliac (internal and external). Common iliac nodes are not regarded as part of the true pelvis and should be considered a distant metastatic (M1) site (see Table 6.2 and Fig. 6.1).

Squamous Cell Carcinoma (SCC)

Squamous cell carcinoma of the bladder is rare in the United States, accounting for only 5% of all bladder cancers. Conversely, in Egypt, where the incidence of infection with *Schistosoma haematobium* is much greater, SCC accounts for approximately 75% of all bladder cancers. There, 18% of all cancers are from the bladder. These lesions are often nodular and fungating, usually confined to the bladder, but urethral or ureteral involvement has been described in up to half of the cases. Keratin pearls are characteristic upon microscopic examination. Cystectomy affords the best long-term survival for patients with squamous cell cancer.

Table 6.2. Bladder cancer TNM staging

Primary Tumor (T)

Tx	Primary tumor cannot be assessed
T0	No evidence of primary tumor
Ta	Noninvasive papillary carcinoma
Tis	Carcinoma in situ: "flat tumor"
T1	Tumor invades subepithelial connective tissue
T2	Tumor invades muscle
T2a	Tumor invades superficial muscle (inner half)
T2b	Tumor invades deep muscle (outer half)
T3	Tumor invades perivesical tissue
T3a	Tumor invades microscopically
T3b	Tumor invades macroscopically (extravesical mass)
T4	Tumor invades contiguous structures/organs (prostate, vagina, uterus, pelvic wall, abdominal wall)
T4a	Tumor invades uterus, vagina, or prostate
T4b	Tumor invades pelvic and/or abdominal wall

6

Regional Lymph Nodes (N)

Nx	Regional lymph nodes cannot be assessed
N0	No regional lymph node metastasis
N1	Metastasis in single lymph node (<2 cm in greatest dimension)
N2	Metastasis in single or multiple lymph nodes (>2 cm but all <5 cm in greatest dimension)
N3	Metastasis in a lymph node >5 cm in greatest dimension

Distant Metastasis (M)

Mx	Distant metastasis cannot be assessed
M0	No distant metastasis
M1	Distant metastasis present

Stage Grouping of Bladder Cancer Stage

0a	Ta	N0	M0
0is	Tis	N0	M0
I	T1	N0	M0
II	T2a	N0	M0
	T2b	N0	M0
III	T3a	N0	M0
	T3b	N0	M0
	T4a	N0	M0
IV	T4b	N0	M0
	any T	N1-3	M0
	any T	any N	M1

Histopathologic Grading of Bladder Cancer

Grade

Gx	grade cannot be assessed
G1	well differentiated tumors
G2	moderately differentiated tumors
G3-4	poorly differentiated or undifferentiated tumors

Adapted with permission from the American Joint Committee on Cancer (AJCC®), Chicago, IL. From AJCC® Cancer Staging Manual, 5th ed. Philadelphia: Lippincott-Raven Publishers, 1997.

Other Bladder Cancers

Adenocarcinoma of the bladder is rare, occurring in <2% of bladder tumors, and can be primary, urachal, or metastatic. Adenocarcinoma can occur in urinary conduits (ileal conduits, ureterosigmoidostomies), and in association with bladder augmentation. Urachal carcinomas are rare, usually adenocarcinoma, and have a poor prognosis. Carcinoma metastatic to bladder is also rare and can include tumors of the rectum, breast, stomach, endometrium, prostate, and ovary.

Clinical Presentation

Symptoms

Most patients with bladder cancer (>80%) will present with total gross painless hematuria. One caveat needs be stated—patients with CIS can present with irritative voiding symptoms, but this is almost always in the presence of microhematuria. Symptoms of more advanced disease can include weight loss, malaise, anorexia, flank pain (ureteral obstruction), bone pain, pelvic pain, and lower extremity edema.

Physical Examination

Most patients with bladder cancer will have a normal physical examination. Those patients with T3 or greater disease where a mass can be appreciated or have signs of metastatic disease are an exception to this rule.

Urine Cytology

Sometimes malignant cells can be visualized from spun urine in patients with bladder cancer. Papanicolaou first described this in 1945. Although urine cytologies are routinely collected when evaluating a patient with hematuria or a suspected urothelial tumor, the sensitivity of the test is low and there is much diagnostic variation among cytopathologists. Many cytologies will be falsely negative, particularly in patients with well-differentiated bladder tumors. Collection of three cytologies on consecutive days has increased the sensitivity of this test from approximately 40-60%. Accuracy is improved when specimens are obtained from bladder washings, since more cellular elements will be obtained when compared to voided specimens. The collection of urinary cytologies, despite its limitations, is typically used in high-risk populations and for surveillance in patients with previously diagnosed urothelial cancers.

Other Urine Markers

The BTA (bladder tumor antigen) test which uses IgG-coated particles has in most studies been shown to be inferior to urine cytologies in the diagnosis of superficial bladder cancer. Other studies evaluating urinary nuclear matrix proteins (NMP22 test), fibrin degradation products (FDP assay), telomerase activity, and hyaluronidase levels have similarly concluded these tests to be inferior to the current gold standard for cytologic screening, urine cytologies.

Evaluation

Cystoscopy with biopsy is the gold standard in the diagnosis of a bladder tumor. Excretory urography (IVP or intravenous pyelogram) is used to evaluate the urothelium in patients with hematuria or suspected lesions. Bladder tumors can be seen with this modality as filling defects within the bladder or as a mucosal irregularity; however, it is not a sensitive nor specific test in the diagnosis of bladder cancer.

Initial workup for patients with microhematuria or suspicion for a urothelial tumor should include IVP, three urine cytology specimens, and cystoscopy. Those found to have a bladder lesion should undergo cystoscopy under anesthesia, bimanual examination, biopsy of lesion (for tissue diagnosis) and complete removal of visualized tumor. Random bladder biopsies and biopsies of the prostatic fossa are also performed to assess the remainder of the "normal appearing" urothelium for possible synchronous microscopic disease. Transurethral resection (TURBT), fulguration, or laser ablation can accomplish tumor eradication. Bladder perforation can occur as a complication of transurethral resection, which can be intraperitoneal if tumor resection is performed at the dome of the bladder. Small bowel injury and subsequent peritonitis have been reported as a rare complication of laser ablation.

Extensive radiographic and laboratory workup (liver function tests, CT scan, MRI, chest x-ray, bone scan), although controversial, is not the standard of care for patients found to have low-grade, superficial bladder tumors. If there is a suspicion of distant or locally advanced disease, a more thorough evaluation is indicated (see Figs. 6.2-6.5).

Intravesical Agents

Indications for the use of intravesical topical therapy in the treatment of superficial bladder cancer include disease that cannot be adequately treated endoscopically (CIS), to prevent disease recurrence, and to assist possibly in the prevention of progression of disease.

BCG (Bacillus Calmette-Guerin) is a live attenuated vaccine and requires an immunocompetent host for success. BCG is a nonspecific stimulant of the immune system (immunotherapy) which exerts its effects through T lymphocytes, B lymphocytes, macrophages, and activated killer cells. BCG has been shown to diminish the incidence of superficial bladder tumor recurrence, disease progression, cystectomy rates, and mortality.

Although a six-week course is standard therapy, many practitioners feel this is insufficient. Several investigators have shown a second six-week course improves the success of treatment. Maintenance BCG is controversial and has not been uniformly proven to be superior to standard induction therapy. Lamm et al have suggested a "6+3" regimen. Patients receive standard induction therapy for six weeks then three weekly installations at three months, six months, and every six months thereafter for three years. The Southwest Oncology Group Study found BCG to both improve disease recurrence and progression to muscle invasive disease.

Approximately 80% of patients will experience some lower urinary symptoms after BCG treatment, primarily cystitis/dysuria. Low-grade fever (<38.5°C) occurs in about 30% of patients. Both of these usually resolve within 48 hours and do not require any treatment. Other complications are less common and include granulomatous prostatitis or epididymitis, ureteritis and ureteric obstruction, nephritis, pneumonitis, hepatitis, allergic reactions, and BCG sepsis. Evaluation and treatment is warranted for those patients presenting with more severe complications, fever, and/or persistent symptoms.

6

6

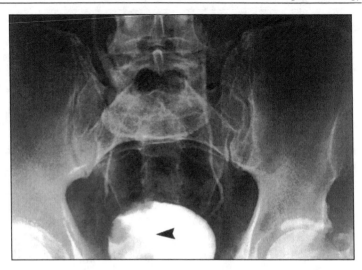

Fig. 6.2. Seventy-one-year-old man who presented with total gross painless hematuria. IVP shows irregular filling defect on the right side of the bladder (black arrow).

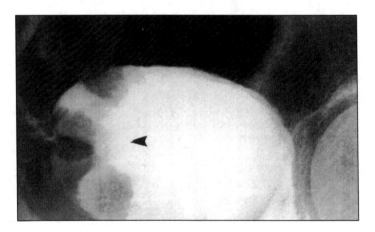

Fig. 6.3. Magnified view of bladder that demonstrates multiple filling defects on the right side of the bladder wall (black arrow). Cystoscopy and transurethral resection revealed a papillary transitional cell bladder cancer.

BCG is the most effective intravesical therapy for superficial bladder cancer, particularly for CIS. Further studies are required to determine the optimal dose and treatment regimen. There is one situation where chemotherapy holds an advantage—a single dose of intravesical chemotherapy immediately after or within 24 hours of tumor resection does diminish the incidence of tumor recurrence.

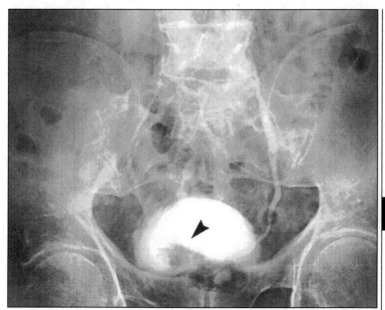

Fig. 6.4. Sixty-four-year-old asymptomatic male smoker who was noted to have microhematuria on urinalysis. IVP shows large filling defect at the bladder base (black arrow) and bowel gas overlying a portion of the bladder. Cystoscopy and transurethral resection revealed papillary transitional cell bladder cancer.

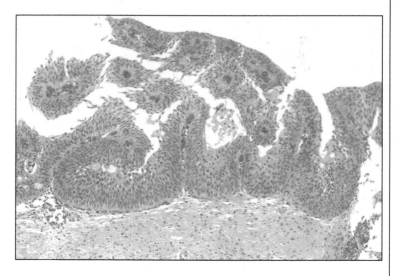

Fig. 6.5. Low-grade, Ta papillary bladder transitional cell carcinoma (40X magnification).

6

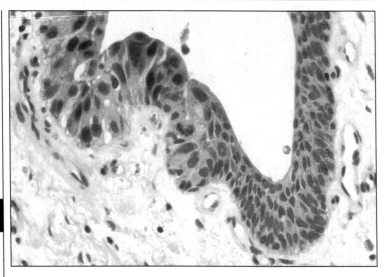

Fig. 6.6. Carcinoma in situ (40X magnification).

Unfortunately, there is no conclusive evidence to support that intravesical chemotherapy decreases the incidence of disease progression. Intravesical immunotherapy must not be given immediately after resection or in the presence of gross hematuria or infection due to the risk of BCG sepsis and the above mentioned complications. All adjuvant intravesical therapies decrease the probability of disease recurrence when compared to resection alone. Currently, BCG is the best intravesical agent for most patients (see Table 6.3).

Treatment

Non-muscle invasive bladder cancer (stages Ta, T1, and Tis): The bladder cancer clinical guidelines panel (1999) convened and made recommendations for patients with non-muscle invasive bladder cancer (Stages Ta, T1, and Tis). The results were based upon a meta-analysis of the literature of outcomes data from articles pertaining to non-muscle invasive bladder cancer from 1966 to 1998. Three examples of patients were described to assist in guiding management.

Sample patient one is noted to have an abnormal growth on the urothelium but has not been diagnosed with bladder cancer. This patient, although likely to have a urothelial cancer, should have histologic confirmation of cancer before treatment is initiated. It is important to obtain pathologic confirmation by biopsy before treatments are begun which would not yield tissue (e.g., fulguration or intravesical therapy).

Sample patient two has biopsy proven bladder cancer (Ta, T1, or Tis) of any grade and has not had intravesical therapy. Consensus treatment for this patient is removal of all visible tumor by either transurethral resection, fulguration, or with laser. Low-grade Ta tumors do not uniformly need adjuvant intravesical therapy as there is a low rate of disease progression, but is an option for select patients. For

Table 6.3. Intravesical agents other than BCG

Agent	Mechanism	Side Effects
1. Interferons	Mediate host immune response	Flu-like symptoms
2. Keyhole Limpet Hemocyanin (KLH)	Immune stimulator (antigenic pigment from the mollusc *Megathuria crenulate*)	No major toxicity
3. Mitomycin C	Inhibition of DNA synthesis	Cystitis, rash, dec bladder capacity
4. Thiotepa (triethylenethio-phosphramide)	Alkylating agent	Myelosuppression
5. Doxorubicin (adriamycin)	Inactivates DNA topoisomerase II	Cystitis
6. Valrubicin	Interferes with DNA topoisomerase II	Cystitis
7. Photodynamic therapy	IV administration of photosensitizer is activated by laser light, damaging cells	Cystitis, bladder contracture, skin photosensitivity reactions

6

patients with biopsy proven CIS or after resection of T1 or high-grade Ta tumors, adjuvant intravesical BCG or mitomycin C is recommended. These agents have been found to be more efficacious than Thiotepa or doxorubicin in reducing recurrence. A higher risk of progression to muscle invasive disease is seen in those patients with large tumor volume, high-grade tumor, diffuse disease, prostatic urethral involvement, and incomplete resection (due to tumor location at inaccessible site). Some patients with CIS or T1 disease may opt for more aggressive therapy, such as cystectomy.

Sample patient three has histologically proven CIS or high-grade T1 disease after one course of intravesical immunotherapy or chemotherapy. Although the risk to muscle invasive disease is higher in this patient, options include a second course of intravesical therapy or cystectomy.

The panel concluded that all intravesical agents when used as adjuvant therapy improved rates of recurrence of bladder cancer when compared to transurethral resection alone. Nevertheless, there was no conclusive evidence that intravesical chemotherapy altered the rate of progression to muscle invasive disease. Although there is no standard algorithm, the upper urinary tract should be evaluated periodically, particularly in patients with CIS, multiple, or recurrent tumors who are more likely to have upper tract disease. These issues should be discussed with the patient and should assist in guiding management.

Other studies have shown high-dose vitamins (A, B_6, C, and E) to reduce tumor recurrence and can be considered in patients who wish to pursue this adjunctive treatment.

Cystectomy

Radical cystectomy with regional lymphadenectomy is the gold standard for treatment of muscle invasive bladder cancer. Relapse is a function of depth of invasion of the primary tumor, involvement of other organs and lymph node status. Of patients with high-grade/stage (T3 or greater) disease, 50% will have positive lymph nodes

at time of cystectomy. The five- year survival can be as high as 50% for those patients with N1 disease and 20% for more significant nodal involvement. For tumors
that are locally advanced/invasive, cystectomy cures <10%. After cystectomy, approximately half of the patients will progress to metastatic disease, usually within
two years. Of these patients, two-thirds relapse at a distant site and one-third relapse
in the pelvis. Compared to other options, cystectomy provides the best chance for
long-term survival in select patients with muscle invasive disease. Urethrectomy should
be performed if there is gross evidence of tumor in the anterior urethra, or if there is
tumor extension into the prostatic urethra or stroma.

Urinary Diversion

Although beyond the scope of this discussion, urinary diversion is performed
both with radical cystectomy for definitive treatment or for palliation. Intestinal
segments (large and small bowel) can be used, each with particular advantages, disadvantages, and metabolic sequelae. They can be orthotopic, heterotopic, continent, or incontinent.

Partial Cystectomy

Many patients wish to preserve their native bladders and avoid radical surgery.
Partial cystectomy is an option for a select group of patients. A recent review considered patients candidates for partial cystectomy if the following conditions were met—
single tumor in upper portion of bladder, absence of multifocal disease or CIS,
normal upper tracts, and absence of metastatic disease. The relapse rate in these
patients is high (up to 70%), and they still need close surveillance. Neoadjuvant
chemotherapy has been given to select patients prior to partial cystectomy. Positive
prognostic factors included small tumor, no hydronephrosis, papillary tumor, and
complete response from induction chemotherapy. A recent review evaluated the overall
success of bladder preservation and compared it to total cystectomy. The conclusion
was that the risk of recurrence and progression coupled with the delay of the elimi-
nation of cancer supports radical cystectomy over bladder preservation. A recent
review evaluating the correlation between TURBT and cystectomy specimens found
that a significant number (52%) of patients were under-staged by TURBT. The
conclusion drawn from this was that staging from initial resection specimens needs
to improve overall staging and assist in stratification of patients for recurrence and
progression. Patients who feel strongly about retaining their native bladders may be
considered for partial cystectomy, informed of the high risk of relapse and possible
progression, and counseled that radical cystectomy is the standard of care and likely
provides the greatest opportunity for long-term cure.

Radiation

Although radical cystectomy has become the gold standard for muscle invasive
bladder cancer, radiotherapy has been used as a first-line treatment in some centers
in Europe. Preoperative radiation therapy has not been shown to improve survival
in patients who undergo radical cystectomy. Although newer techniques with radiation dose, fractionation, and field are promising, radiotherapy should be considered
for patients who are unable to undergo cystectomy or for those who desire alternative treatments.

Chemotherapy

Chemotherapeutic regimens are highly variable for the treatment of bladder cancer. The standard therapy is M-VAC (methotrexate, vinblastine, doxorubicin, and cisplatin). Partial response rates between 40-70% have been proposed for M-VAC as neoadjuvant or adjuvant therapy after cystectomy. The goal of chemotherapy is to destroy micrometastatic disease that may be present at diagnosis as well as possible downstaging of the primary tumor. Disadvantages of chemotherapeutic regimens include toxicities and delay of local definitive therapy. Most trials have not demonstrated a survival difference. Some studies have shown a marginal survival and progression benefit, particularly those patients who responded and were down-staged with neoadjuvant chemotherapy. Skinner et al demonstrated a survival advantage in patients who received adjuvant chemotherapy, particularly those patients with no nodal disease. A recent review noted that perioperative chemotherapy improved relapse-free survival and pelvic failure, but not metastatic progression in high risk patients (>T3, tumor >3 cm, creatinine >1.5X normal).

Follow-Up

Although there is some variation in the algorithm across institutions and practitioners, follow-up for patients with superficial bladder cancer (after definitive local treatment) is surveillance cystoscopy and urinary cytology every three months for one year (after the initial treatment), then surveillance every four months for one year, every six months for one year, and annually thereafter. If recurrence is detected, the cycle should restart from the beginning of the algorithm after local retreatment. Surveillance of the upper urinary tracts is less clear. At our institution, the upper tracts are imaged (either with IVP, ultrasound, or retrograde pyelography) approximately every 3-5 years, but more frequently for patients with carcinoma in situ, multifocal disease, or recurrent tumors.

The Future

Bladder cancer is a complex and rapidly evolving subject. Research investigating the role of biological markers (p53, p21, and Ki-67 antigen) in superficial bladder cancer is ongoing. Trials further investigating intravesical interferons and photodynamic therapy are continuing. Other studies using chemotherapeutic agents such as gemcitabine and the taxanes (paclitaxel and docetaxel) have shown some promise and continue to be studied. One can be hopeful that the future will provide improved methods for diagnosis, screening, and treatment for bladder cancer.

References

1. Messing EM, Catalona W. Urothelial tumors of the urinary tract. In: Walsh PC et al., eds. Campbells Urology. 7th ed. Philadelphia: WB Saunders Co., 1998:2327-2383.
2. Lamm DL, Blumenstein B, Sarasody M et al. Significant long-term patient benefit with BCG maintenance therapy: a Southwest Oncology Group Study. J Urol 1997; 157(4 Suppl):213.
3. Rischmann P, Desgrandchamps F, Malavaud B et al.. BCG intravesical instillations: recommendations for side-effects management. Eur Urol 2000; 37(Suppl 1):33-36.

4.   Smith JA Jr, Labasky RF, Cockett AT et al. Bladder cancer clinical guidelines panel summary report on the management of non-muscle invasive bladder cancer (stages Ta, T1 and TIS). The American Urological Association. J Urol 1999; 162(5):1697-1701.

5.   Lamm DL, Riggs DR, Shriver JS et al. Megadose vitamins in bladder cancer: a double-blind clinical trial. J Urol 1994; 151(1):21-26.

6.   Dandekar NP, Tongaonkar HB, Dalal AV et al. Partial cystectomy for invasive bladder cancer. J Surg Oncol 1995; 60(1):24-29.

7.   Herr HW, Scher HI. Neoadjuvant chemotherapy and partial cystectomy for invasive bladder cancer. J Clin Oncol 1994; 12(5):975-980.

8.   Montie JE. Against bladder sparing: surgery. J Urol 1999; 162(2):452-457.

9.   Cheng L, Neumann RM, Weaver AL et al. Grading and staging of bladder carcinoma in transurethral resection specimens. Correlation with 105 matched cystectomy specimens. Am J Clin Pathol 2000; 113(2):275-279.

10.   Sengelov L, von der Maase H. Radiotherapy in bladder cancer. Radiother Oncol 1999; 52(1):1-14.

11.   Ennis RD, Petrylak DP, Singh P et al. The effect of cystectomy, and perioperative methotrexate, vinblastine, doxorubicin and cisplatin chemotherapy on the risk and pattern of relapse in patients with muscle invasive bladder cancer. J Urol 2000; 163(5):1413-1418.

Suggested Readings

General Information

1.   Droller MJ. Advanced bladder cancer. Urol Clin N Am 1992; 19(4):663-744.

2.   Laughlin KR. Superficial bladder cancer: New strategies in diagnosis and treatment. Urol Clin N Am 2000; 27(1):1-197.

3.   Herr HW. Tumor progression and survival of patients with high grade, noninvasive papillary (TaG3) bladder tumors: 15-year outcome. J Urol 2000; 163(1):60-62.

4.   Millan-Rodriguez F, Chechile-Toniolo G, Salvador-Bayarri J et al. Multivariate analysis of the prognostic factors of primary superficial bladder cancer. J Urol 2000; 163(1):73-78.

5.   Droller MJ, Gospodarowicz MK. Staging of bladder cancer. In: Vogelzang N et al, eds. Comprehensive Textbook of Genitourinary Oncology. Baltimore: Williams & Wilkins, 1996:359-370.

6.   Fradet Y. Epidemiology of bladder cancer. In: Vogelzang N et al, eds. Comprehensive Textbook of Genitourinary Oncology. Baltimore: Williams & Wilkins, 1996:298-304.

7.   American Joint Committee on Cancer. Urinary bladder. In: Fleming ID, Cooper JS, Henson DE eds. AJCC Cancer Staging Manual. 5th ed. Philadelphia: Lippincott-Raven, 1997:241-246.

8.   Hermanek P, Hutter RVP, Sobin LH et al, eds. TNM Atlas: illustrated guide to the TNM/pTNM classification of malignant tumors. 4th ed. New York, Berlin: Springer-Verlag Publishers, 1997:310-313.

Intravesical Therapy

1.   Kamat AM, Lamm DL. Intravesical therapy for bladder cancer. Urology 2000; 55(2):161-167.

2.   Losa A, Hurle R, Lembo A. Low dose bacillus Calmette-Guerin for carcinoma in situ of the bladder: long-term results. J Urol 2000; 163(1):68-72.

Surgical Therapy

1.   Shipley WU, Kaufman DS, Heney NM et al. An update of combined modality therapy for patients with muscle invading bladder cancer using selective bladder preservation or cystectomy. J Urol 1999; 162(2):445-451.
2.   Rosário DJ, Becker M, Anderson JB. The changing pattern of mortality and morbidity from radical cystectomy. BJU Int 2000; 85(4):427-430.
3.   Zietman AL, Shipley WU, Kaufman DS. Organ-conserving approaches to muscle-invasive bladder cancer: future alternatives to radical cystectomy. Ann Med 2000; 32(1):34-42.

Chemotherapy/Radiotherapy

1.   Medical Research Council Advanced Bladder Cancer Working Party et al. Neoadjuvant cisplatin, methotrexate, and vinblastine chemotherapy for muscle-invasive bladder cancer: A randomized controlled trial. International collaboration of trialists. Lancet 1999; 354(9178):533-540.
2.   Sternberg CN, Calabro F. Chemotherapy and management of bladder tumors. BJU Int 2000; 85(5):599-610.
3.   Skinner DG, Daniels JR, Russel CA et al. The role of adjuvant chemotherapy following cystectomy for invasive bladder cancer: A prospective comparative trial. J Urol 1991; 145(3):459-464.

6

CHAPTER 7

Urinary Diversion

Marcus L. Quek and John P. Stein

Introduction

Urinary diversion following bladder removal for either benign or malignant disease has undergone significant advances over the past 50 years. The use of bowel in urinary tract reconstruction has added greatly to the urologic surgeon's armamentarium and the treatment of many diseases. The advent of modern surgical technique and instrumentation and advances in anesthetic and perioperative care have led to the continued creativity seen in lower urinary tract reconstruction. The goals of urinary diversion following cystectomy have become more than simply a means to redirect urine and protect the upper urinary tract. Contemporary objectives have emphasized quality of life issues: eliminating the need for external collection devices, cutaneous urostomy stomas, or need for intermittent catheterization; while attempting to maintain a more natural voiding pattern allowing volitional micturition through the intact native urethra. These advances in urinary diversion have provided patients a more normal lifestyle, with an improved quality of life and self-image following cystectomy.

Prior to 1950, ureterosigmoidostomy represented the urinary diversion of choice. As experience with the ureterosigmoidostomy grew, complications of ascending pyelonephritis, electrolyte imbalances, renal deterioration, and secondary malignancies at the ureteral implantation site directed efforts to provide safer and more reliable forms of urinary diversion.

Over the last 50 years, lower urinary tract reconstruction has evolved along three distinct paths: a non-continent cutaneous conduit form of diversion (ileal or colon conduit), a continent cutaneous form of urinary diversion (continent cutaneous), and more recently an orthotopic diversion to the intact, native urethra (neobladder substitute).

In 1950, Bricker introduced the ileal conduit, which would remain the so-called "gold standard" of urinary diversions through the early 1990s. Innovations by investigators, including Gilchrist and Kock, introduced the concept of a continent cutaneous mechanism, thereby obviating the need for a urostomy bag, with the patient performing intermittent catheterization through a continent abdominal wall stoma. Over the past decade, we and others believed that interest in an orthotopic form of diversion has improved patient satisfaction and greater acceptance of radical cystectomy for invasive bladder cancer. In fact, in 1993, at the Fourth International Consensus Conference on Bladder Cancer in Antwerp, Belgium, consensus opinion was that in the properly selected bladder cancer patient, urinary reconstruction to the urethra is the procedure of choice in most centers worldwide.

Urological Oncology, edited by Daniel Nachtsheim. ©2005 Landes Bioscience.

The second major advance in urinary diversion relates to orthotopic reconstruction in female cystectomy patients. Prior to 1990, orthotopic lower urinary tract reconstruction was reserved only for male patients and considered contraindicated in the female subject. Reasons included the fact that the entire urethra was routinely removed during anterior exenteration in women as it was felt necessary to provide an adequate cancer margin. In addition, it was believed that the urethral length in females was inadequate to provide continence for an orthotopic diversion. Based on an extensive pathologic review of female cystectomy specimens removed for transitional cell carcinoma of the bladder, it was shown that the urethra could be safely preserved in the majority of women. These findings have been confirmed in long-term prospective studies and have provided sound pathologic criteria to help identify appropriate female candidates for orthotopic substitution. In addition, elegant neuroanatomic cadaveric dissections of the female pelvis have provided a better understanding of the urethral continence mechanism in women and have showed that women may maintain their continence if the bladder neck and proximal urethra are removed while leaving intact the remaining urethra and rhabdosphincter complex. These important findings have provided a foundation in which to offer women lower urinary tract reconstruction to the urethra.

7

Proper patient selection is critical in determining which form of diversion is appropriate for a patient faced with cystectomy. Attention to detail and surgical technique will ensure the best possible result regardless of the form of urinary diversion chosen.

Conduit Diversions

Despite the current trend toward continent diversions, many patients may best be served with a conduit diversion following cystectomy. The ileal conduit remains a standard to which other forms of diversion can be compared. It is technically simple, quicker to perform than continent urinary reservoirs, has a lower early postoperative complication rate, and has widespread experience in the urologic community. Deciding on which form of diversion to use should be based on several factors including the patient's overall medical condition, the surgical goal (palliation or cure), and the patient's expectations and preferences.

Patients lacking the necessary motivation or inability to understand the possible complications associated with a continent means of diversion, or unwilling to accept the need or responsibility for catheterization, should consider a conduit diversion. Due to the greater intestinal absorption of urine in continent diversions, patients with renal insufficiency (creatinine clearance <35 ml/min or serum creatinine >2.0 ng%) are best served with a conduit. The goal of the surgery, either palliation or cure, also plays a significant role in selecting the type of diversion. In a palliative setting, the easiest and quickest procedure is usually desired, and an ileal conduit should be preferred.

Three basic conduit choices are available—ileal, jejunal, and colonic. Ileal conduits are the most commonly employed owing to their ease of construction and low complication risk. However, it is not advisable to use an ileal segment in patients with extensive prior pelvic irradiation, short bowel syndrome, and regional enteritis. Jejunal conduits are rarely used due to associated electrolyte abnormalities (hyperkalemic, hyponatremic metabolic acidosis). Conduits constructed from colon (transverse, sigmoid, and ileocecal) may also be employed. One advantage of colon conduits is the ability to provide an antireflux mechanism. A non-refluxing ureterocolonic anastomosis can be performed using the tenia coli in a technique described by Leadbetter. Transverse colon may be preferred in patients with extensive pelvic irradiation as this portion of bowel lies outside the field of irradiation. Sigmoid conduits are particularly advantageous when a left-sided stoma is desirable, and in patients undergoing total pelvic exenteration. In this situation, since the patient undergoes a colostomy for fecal diversion, a bowel anastomosis can be avoided. Ileocecal conduits provide a long segment of ileum should an extensive ureteral replacement be required. Contraindications to use of a colonic segment include severe chronic diarrhea and concomitant colonic disease.

Critical to the success of any form of cutaneous diversion (continent or incontinent) is identifying the appropriate cutaneous stoma site. The role of the enterostomal therapist is invaluable in this regard. Evaluation of the patient in the supine, sitting, and standing positions, while avoiding sites overlying scars, creases, belt lines, or bony structures will help ensure a proper fitting external collection appliance. In obese patients, it may be helpful for the patient to wear the urostomy bag for several days preoperatively to ensure satisfactory placement. Generally, the stoma site is located over the lateral aspect of the rectus muscle, thereby helping to prevent parastomal herniation.

As with any form of urinary diversion, strict attention to surgical detail and proper tissue handling is critical to avoid complications and provide the best functional results with the diversion. The ileal conduit is constructed from a segment of distal ileum approximately 15 cm from the ileocecal junction. The authors prefer a mesenteric incision along the avascular plane of Treves between the terminal branches of the ileum and the ileocolonic branches of the superior mesenteric artery. The appropriate length of small bowel is determined (average 15 cm), the proximal end of the ileal limb is closed, and a standard end-to-side spatulated ureteroileal anastomosis performed distal to the closed end. The authors prefer stenting the ureterointestinal anastomosis with the stents exiting the ostomy (Fig. 7.1).

Next, the stoma is created. A small circle of skin is excised at the predetermined site. Care is taken to avoid removing excessive subcutaneous tissue which helps reduce problems of stomal retraction. The anterior rectus fascia is incised in a cruciate fashion large enough to accommodate two fingers. A "rosebud" or Turnbull stoma (author's preference) technique can be performed. Regardless, the loop should be secured to the anterior rectus fascia and should protrude, without tension, at least 2-3 cm above the skin. This facilitates excellent application of the urostomy appliance to the skin without problems of leaking.

Long-term complications following ileal or colon conduits are listed in Tables 1 and 2 and include: pyelonephritis, renal deterioration, ureteral obstruction, development of urinary calculi, and stomal stenosis. Despite these complications, there still remains a role for conduit forms of urinary diversion. The technical ease of the operation coupled with the relatively low short-term morbidity makes conduit diversion attractive for patients. This may be particularly important in those with coexistent medical conditions or with a compromised life expectancy. Regardless, the potential for complications exist and underscore the need for meticulous attention to surgical detail and careful follow-up in these patients.

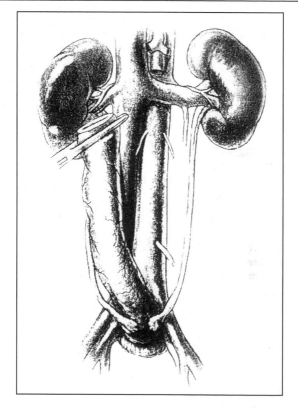

Fig. 7.1. Leadbetter's technique of ileal loop diversion. Ureters are anastomosed end to side near the base of the conduit, and the base is further sutured to the sacral promontory or the overlying fibrous tissue. Reprinted from: Skinner DG, Lieskovsky G, eds. Diagnosis and Management of Genito-Urinary Cancer. 2/e. pp. 635-638. ©1988, with permission from Elsevier.

Principles of Continent Urinary Diversion

In general, patients considered appropriate surgical candidates for radical cystectomy should also be potential candidates for a continent urinary diversion. Relative contraindications relate to the ability of the patient to perform self-catheterization and to care for a neobladder owing to mental or physical impairments. Poor renal and hepatic function should also be considered a contraindication to continent urinary diversion. Advanced age should not be an absolute contraindication as long as the patient maintains the ability for self-care, and an extensive surgical procedure is not precluded by the overall medical condition. Adequate bowel length and an intestinal segment free of inflammatory, neoplastic, or post-irradiated changes should be ensured preoperatively. When the large bowel is to be used for construction of the reservoir, it is prudent to perform a large bowel radiographic study prior to exclude any obvious pathology. This is not required when ileum is used to construct

Table 7.1. *Long term complications with an ileal conduit diversion*

	Number of Patients	Pyelonephritis	Ureteroileal Stenosis	Stones	Stomal Stenosis
Butcher	307	42 (13%)	7 (2%)	10 (3%)	20 (6%)
Johnson	214	33 (15%)	39 (18%)	5 (2%)	11 (5%)
Sullivan	336	65 (19%)	49 (14%)	13 (4%)	17 (5%)
Middleton	90	18 (20%)	9 (10%)	8 (9%)	17 (5%)
Shapiro	90	15 (16%)	20 (22%)	8 (9%)	38 (42%)
Pitts	242	26 (10%)	10 (4%)	14 (5%)	34 (14%)
Totals	**1279**	**199 (16%)**	**134 (10%)**	**58 (4.5%)**	**154 (12%)**

the reservoir. In obese individuals, an orthotopic diversion may be preferred because of the potential difficulty in maintaining an ostomy appliance, as well as the potential difficulties in negotiating a thick abdominal wall while self-catheterizing a continent cutaneous reservoir.

The Reservoir

The "ideal reservoir" should be compliant and accommodate a large volume under low pressure, without reflux or absorption of urinary constituents. The concept of detubularization and folding of the bowel into a spherical form greatly increases the storage capacity with significantly lower internal filling pressures without coordinated peristaltic contractions. The lower intraluminal pressures within the reservoir should allow for improved continence and decreased pressure on the upper urinary tracts. Important principles of reservoir construction include: configuration—which determines geometric capacity (volume = height x radius^2), accommodation—which relates pressure and volume to mural tension (Law of Laplace, tension = pressure x radius), and compliance—which relates to the physical characteristics of the particular bowel segment.

Virtually every segment of the gastrointestinal tract has been utilized in urinary reservoir construction. The choice depends on several factors including: concomitant bowel disease, metabolic consequences, and surgeon preference. The following describes the metabolic considerations and gastrointestinal consequences associated with the various intestinal segments used in urinary reconstruction.

Ileum

When an ileal segment is utilized in urinary tract reconstruction, potential metabolic derangements include a hyperchloremic hypokalemic metabolic acidosis and osteomalacia due to hypercalciuria. In those with normal renal function, these effects are minimal. Megaloblastic anemia is also a potential complication due to insufficient vitamin B12 absorption at the terminal ileum. Bile salt malabsorption syndromes and osmotic diarrhea are also potential problems.

Jejunum

Use of jejunum in urinary diversion is rare due to a severe "jejunal conduit syndrome" characterized by hypovolemia, hyperkalemia, hyponatremia, and hypochloremic metabolic acidosis. These metabolic abnormalities can occur even with minor degrees of renal insufficiency.

Colon

Similar to ileal segments, colonic reservoirs may be subject to hyperchloremic hypokalemic metabolic acidosis, hypomagnesemia and hypocalcemia. Diarrhea and malabsorption syndromes may also occur. Another long-term complication of colonic diversions relates to the potential for tumor formation in the colonic segment. The lifetime risk of tumor formation has been estimated to be about 5 to 10%. There is a strong predilection for the ureteral anastomotic site in ureterosigmoidostomy, in particular, and is typically an adenocarcinoma.

Stomach

As the stomach continues to excrete acid, a hypochloremic metabolic alkalosis may ensue, especially in those with impaired renal function. In terms of gastrointestinal function, a dumping syndrome may result as well as megaloblastic anemia due to decreased intrinsic factor secretion and loss of vitamin B_{12} absorption.

The reconstructive surgeon must be familiar with the various options for urinary diversion and adept at using the various bowel segments depending on the particular surgical situation and the medical condition of the patient. The principle of detubularization allows for a low pressure, high volume reservoir.

7

Continence Mechanisms

Various methods for achieving continence have been described in the literature. Continence in urinary diversion is dependent on the relationship between the outlet resistance and internal reservoir pressure. The continence mechanism must withstand increases in intra-reservoir pressure not only at rest but also during changes in position or under Valsalva pressure. Despite the wide variety of efferent continence mechanisms, the ideal mechanism has yet to be defined. The most commonly applied techniques available for continent cutaneous diversions rely on either intraluminal mechanisms (intussuscepted nipple valve or flap valve) or an extraluminal mechanism (tapered or plicated efferent limb). Alternatively, patients undergoing an orthotopic reconstruction rely on the inherent properties of the rhabdosphincteric complex.

The intussuscepted nipple valve continence mechanism, as exemplified in the Kock pouch, is based on the physiologic principle that the filling pressure inside the reservoir will be simultaneously applied to the outside of the intraluminal nipple, thereby causing compression of the nipple and prevention of leakage. Concern with this continence mechanism developed when it was realized that as the urinary reservoir fills, the increasing intraluminal pressure is applied not only to the nipple, but also to the wall of the reservoir laterally. This results in distraction and effacement of the nipple and potential compromise of the continence mechanism. Modifications of this technique have been proposed in which the nipple valve is stabilized to the wall of the reservoir; however, technical issues with regard to construction and complications of the intussuscepted nipple valve have lessened the enthusiasm for this continence mechanism.

The flap valve principle relies on compression of the lumen of a catheterizable channel that is fixed and tunneled along the inner wall of the reservoir. As the reservoir fills, the channel is compressed, thus preventing incontinence. This principle is analogous to the techniques described for ureteral reimplantation for vesicoureteral reflux. As with ureteroneocystostomy, a 5 to 1 ratio of intramural tunnel length to channel lumen diameter has been advocated to ensure dependable continence. When the appendix is used in the flap valve mechanism it has been coined the "Mitrofanoff principle" after Mitrofanoff who described the use of a continent catheterizable appendicovesicostomy in children with neurogenic bladders. Other structures have been employed as the catheterizable efferent limb including: ureter, fallopian tube, vas deferens, tubularized bladder wall, tubularized stomach, and tapered ileum.

An innovative application of the flap valve mechanism is found in the serous-lined tunnel described by Stein et al in which the efferent limb of ileum is laid in a serosal lined ileal trough formed by the ileal segments of the pouch. The ileal segments are opened in such a way as to create wide flaps of ileum, which may be brought over the tapered ileal segment thereby forming the flap valve mechanism ("T-mechanism"). This has been incorporated as both a continence and antireflux mechanism in the "T-pouch" ileal reservoir.

The extraluminal continence mechanism, as exemplified by the ileocecal reservoirs (Indiana and Florida pouches), is based on a tapered ileal segment with plication of the ileocecal valve for continence. In terms of Laplace's Law, ($T = P \times r$; where T is the mural wall tension, P is the intraluminal pressure, and r is the radius of the lumen), as the radius of the catheterizable channel is decreased, the luminal pressure will subsequently increase ($P = T/r$). Simply stated, the narrower the efferent catheterizable limb, the greater the resistance to urinary leakage. Based on this theoretical premise and confirmed by urodynamic testing, the closing pressure in this extraluminal valve is a constant (depending on the radius of the efferent limb) at any pouch volume. However, when examining the intraluminal mechanism (nipple, flap valve, serous-lined tunnel), the closing pressures proportionally increase as the pouch fills and compresses the valve. Therefore, the intraluminal continence mechanisms may be more reliable and maintain better continence at higher volumes and pressures compared to extraluminal mechanisms.

Orthotopic lower urinary tract reconstruction depends on the patients' inherent intact striated rhabdosphincter complex and corresponding innervation to maintain continence. Preservation of the pudendal nerve supply to the rhabdosphincter complex during cystectomy is thought to be critical in the maintenance of this continence mechanism. The complex musculofascial urethral support system may also play a role in the overall continence of patients undergoing orthotopic reconstruction. As experience with orthotopic diversion increases, so too will the understanding of the multiple factors needed for optimal urinary continence.

Prevention of Urinary Reflux

The reflux of urinary constituents following diversion should be minimized or prevented in order to preserve renal function and prevent pyelonephritis. Problems with urinary reflux first became clinically evident in patients with ureterosigmoidostomy and then later with ileal conduit diversions.

Reflux prevention is clearly important for cutaneous urinary diversions which require intermittent catheterization, as invariably the urine is chronically colonized by bacteria. However, with the increasing popularity of orthotopic diversions, in which the urine is potentially sterile, the issue of reflux prevention is debatable.

Many ureteroenteric antireflux techniques have been described including:

1. a submucosal tunneled anastomosis in the wall of a colonic reservoir (Goodwin, Leadbetter);
2. an intussuscepted nipple valve (Kock);
3. an interposed long isoperistaltic segment of ileum (Studer);
4. LeDuc mucosal groove ureteroileal anastomosis; and
5. the serosal-lined tunnel of the T-mechanism.

The principles of the submucosal tunneled ureterocolonic anastomosis and intussuscepted nipple valve are identical to those outlined for intraluminal continence mechanisms described previously. The ureters drain into the reservoir through either a tunnel of colonic submucosa or a one-way nipple valve, both of which will compress as the reservoir fills, thereby preventing reflux.

Studer and associates advocate a ureteral anastomosis into a long (20 cm) dynamic isoperistaltic ileal segment that is connected to a low pressure orthotopic neobladder reservoir. Proponents of the so-called Studer pouch argue that reflux of sterile urine from a low pressure reservoir may not have any clinical consequence and that the complications of late stenosis from various antireflux techniques outweigh their theoretical advantage of protecting the upper urinary tract.

Others strongly argue that antireflux procedures are critical to preserve renal function and are important to all forms of lower urinary tract reconstruction, even those undergoing orthotopic diversion. Reasons include the fact that patients, even with an orthotopic neobladder, may have colonized bacteriuria. This may be even more important as the overall treatment (medical and surgical therapies) for pelvic malignancies improves and patients live longer following exenteration and urinary diversion. Furthermore, several groups (Abol-Enein and Ghoneim and Stein and Skinner) have developed novel antireflux mechanisms intended to improve on existing techniques and eliminate or reduce the complications associated with previous antireflux mechanisms. Longer follow-up will obviously be required to accurately evaluate the isoperistaltic Studer limb, as well as these more recent antireflux techniques to determine if reflux prevention in patients undergoing orthotopic diversion is truly necessary.

Continent Cutaneous Reservoirs

Kock's introduction of a continent ileal diversion in 1982 reintroduced enthusiasm in a continent cutaneous form of urinary diversion. Many continent cutaneous urinary diversions have subsequently been described. Currently, it does not appear that any one reservoir construction is clearly better or preferable to another. The orthotopic neobladder has gained recent popularity and has become the primary form of reconstruction in most patients requiring urinary diversion. Despite this enthusiasm with orthotopic diversion, there still remains definite indications for performing a continent cutaneous diversion. Patients with tumor involvement of the proximal urethra (distal surgical margin) determined at the time of surgery on frozen section analysis, or those patients without an intact, functional rhabdosphincter mechanism are best served with a cutaneous form of diversion.

The Indiana (ileocecal) pouch remains one the most popular continent cutaneous urinary diversions because of its technical simplicity and ease of construction. This reservoir is constructed from terminal ileum (efferent limb), cecum and ascending

Table 7.2. Long term complications with a colon conduit diversion

	Number of Patients	Pyelonephritis	Ureteroileal Stenosis	Stones	Stomal Stenosis
Morales	46	8 (17%)	6 (13%)	2 (4%)	6 (13%)
Althausen	70	5 (7%)	6 (8%)	3 (4%)	2 (2%)
Totals	**116**	**13 (11%)**	**12 (10%)**	**5 (4%)**	**8 (7%)**

colon (reservoir). The colon is detubularized about three fourths of its length and the ureters are anastomosed through a tunnel along the posterior colonic tenia. The ascending colon is folded down to the opened cecum, and the continence mechanism is created by plication of the distal aspect of the terminal ileum which reinforces the ileocecal valve. The efferent limb and continence mechanism have been improved by using a stapling technique to taper the efferent limb prior to suture plicating the ileocecal valve. Advantages with the Indiana pouch included a low complication rate and a short learning curve allowing it to be safely performed by surgeons who perform a limited number of continent cutaneous diversions (Fig. 7.2).

The Mainz pouch, first described in 1985, incorporates cecum, ascending colon and terminal ileum for construction of the reservoir. The operative technique has been modified to include the intact ileocecal valve as a means to further stabilize the intussuscepted ileal limb. The colon and 30 cm of adjacent terminal ileum are detubularized and re-approximated side-to-side to form the reservoir portion of the pouch. An intact non-detubularized portion of proximal ileum is then intussuscepted and stapled through the intact ileocecal valve for the continence mechanism. The ureters are then placed into a submucosal tunnel along the posterior tenia of the colon.

The appendix has also been incorporated as a continence mechanism in lower urinary tract reconstruction. The application of the Mitrofanoff principle requires a normal appendix (approximately 7 cm of length) without inflammatory changes and should accommodate at least a 14 F catheter. Duckett and Snyder first incorporated the appendix in the so-called "Penn Pouch." They report excising the appendix with a button of cecum, reversing it upon itself, and performing a tunneled reimplant into the tenia of the detubularized colon. A modification of this Mitrofanoff principle was subsequently applied in the Mainz configuration of a right colon reservoir. In the Mainz procedure, the ascending colon and terminal ileum are used to form the reservoir portion of the pouch, and the ureters are tunneled into the posterior tenia of the colon. The intact appendix is placed into a wide tunnel created in the tenia extending 5 or 6 cm from the base of the appendix. Mesenteric windows are created in the mesoappendix and the appendix is then folded cephalad into the tunnel. Seromuscular sutures are placed through the opened mesoappendix windows to anchor the appendix and create a flap-valve technique with preservation of the appendicular blood supply. A variation of this diversion has been performed at the University of Southern California using the ascending colon for the reservoir portion of the pouch and a small portion of intact terminal ileum for the afferent segment. The appendix is similarly prepared as in the Mainz pouch; however, the antireflux mechanism utilizes the intact ileocecal valve (reinforced if necessary) with an end-to-side ureteroileal anastomosis performed to the proximal, intact terminal ileum.

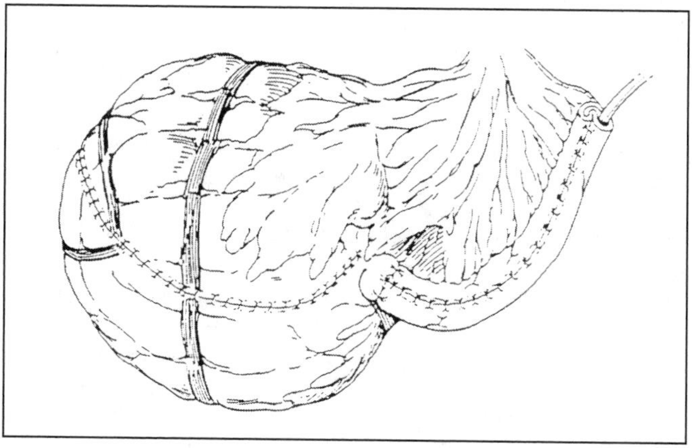

Fig. 7.2. Indiana Pouch. Note tapering of illeal segment over catheter and detubularization of right colon pouch. (Reprinted with permission from: Ahlering TE et al. J Urol 1991; 145:1156-1158.)

7

Orthotopic Urinary Diversion

Although the ideal bladder substitute remains to be developed, the orthotopic neobladder most closely resembles the original bladder in both location and function. A natural extension of the continent cutaneous urinary diversion is the orthotopic neobladder anastomosed directly to the native intact urethra. The orthotopic neobladder eliminates the need for a cutaneous stoma and a cutaneous collection device. This form of lower urinary tract reconstruction relies on the intact rhabdosphincter continence mechanism; eliminating the need for intermittent catheterization and the often plagued efferent continence mechanism of most continent cutaneous reservoirs. Continence is maintained by the external striated sphincter muscle (rhabdosphincter muscle) of the pelvic floor, while voiding is accomplished by concomitantly increasing intraabdominal pressure (Valsalva), with relaxation of the pelvic floor. The majority of patients undergoing orthotopic reconstruction are continent and void to completion without the need for intermittent catheterization.

Camey's pioneering work with orthotopic reconstruction in carefully selected male patients has subsequently evolved into a common form of lower urinary tract reconstruction in all patients requiring urinary diversion. Two important criteria must be fulfilled when considering any patient for orthotopic urinary diversion. First, under no circumstance must the cancer operation be compromised by the reconstruction at the ureteroenteric anastomosis, retained urethra, or surgical margins. Second, the rhabdosphincter mechanism must remain intact to provide a continent means of storing urine. If these criteria can safely be maintained, the patient may be considered an appropriate candidate for orthotopic urinary diversion.

Orthotopic urinary diversion was initially performed in selected male subjects and was not considered technically possible in females. Total urethrectomy was routinely performed during cystectomy for women. In addition, a lack of understanding of the continence mechanism in women also dampened the enthusiasm of orthotopic reconstruction in females. With a better understanding of the continence mechanism in women and discovering that the urethra can safely be preserved in selected female patients undergoing radical cystectomy; orthotopic lower urinary tract reconstruction has now become a viable option in women.

The literature is replete with various forms of neobladder reconstruction, utilizing different portions of bowel, with and without various types of antireflux techniques. The Hautmann ileal neobladder is constructed from 60-80 cm of detubularized ileum fashioned into a W plate. The ureters are reimplanted into the lateral segments of the W using the Le Duc antireflux technique. Studer and associates have advocated a modification of the ileal neobladder similar to the cup patch technique described by Goodwin in 1959. This form of diversion incorporates a long isoperistaltic afferent limb; by definition this is not an antireflux mechanism. Intermediate follow-up with this form of diversion has been promising.

Abol-Enein and Ghoneim have described a neobladder with a novel antireflux technique incorporating an uretero-ileal implantation technique via a serous-lined extramural tunnel. This technique was successfully applied in an ileal neobladder arranged in a W-shape configuration. This technique was advocated because of its technical simplicity and excellent functional results.

The orthotopic Kock ileal reservoir has also been a popular form of neobladder reconstruction. With an extensive review of their experience with the Kock ileal neobladder, Stein and Skinner reported approximately a 10% complication rate associated with the intussuscepted antireflux nipple in over 800 patients followed for an average of 6 years. The most common complications associated with the intussuscepted afferent nipple include: the formation of calculi (usually on exposed staples that secure the afferent valve) in 5%, afferent nipple stenosis (thought to be caused by the ischemic changes resulting from the mesenteric stripping required to maintain the intussuscepted limb) in 4%, and extussusception (prolapse of the afferent limb) in 1% of patients. Although the majority of these afferent nipple valve complications (60%) can be easily managed with endoscopic techniques on an outpatient basis, they nonetheless, may result in some morbidity. In fact, approximately 3% of all patients undergoing a continent Kock ileal reservoir will require an open surgical revision to repair an afferent nipple complication.

The need to improve upon the antireflux mechanism to reduce or eliminate the observed late complications of the afferent intussuscepted nipple valve became obvious. Based on reports from Ghoneim's group using the ureteral extra-serosal tunnel, as well as our own experience with the Mitrofanoff appendiceal subserosal tunnel, we described a novel antireflux technique using an afferent ileal segment that is permanently anchored within a serosal-lined tunnel with complete preservation of the mesentery (T-pouch). The unique aspect of the T-pouch is in the maintenance of the vascular arcades of the distal afferent ileal segment by opening the Windows of Deaver. This afferent ileal segment is fixed within the serosal trough of the ileal reservoir while maintaining a well-vascularized afferent limb to prevent problems with stenosis. Furthermore, no exposed staples exist in this reservoir, which should eliminate problems with stone formation. The early clinical and functional results with this urinary diversion have been excellent (Fig. 7.3).

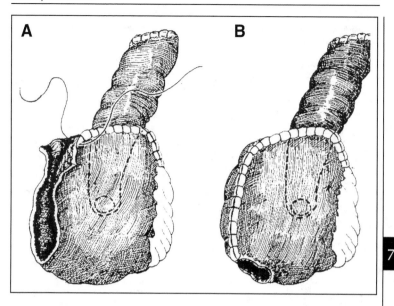

Fig. 7.3. A) Anterior suture line is completed with two layers of continuous 3-0 polyglycolic acid suture. Anterior suture line is stopped just short of right side to allow insertion of index finger, forming neourethra. B) Completion of T-pouch. Most mobile and dependent portion of reservoir is anastomosed to urethra after ureteroileal anastomosis. (Reprinted with permission from: Stein. J Urol 1998; 159:1836-1842.)

Conclusion

The evolution of urinary diversion has been a remarkable process over the past 150 years. It is well accepted that the creation of a continent urinary reservoir requires three major components: (1) a large capacity, low pressure reservoir; (2) an effective continence mechanism; and (3) an effective, permanent antireflux mechanism. However, the search for the perfect urinary diversion remains to be found. It will only be the motivated and thoughtful surgeon, improving on existing concepts and techniques, who becomes closer to the ideal form of lower urinary tract reconstruction.

Selected Readings

1. Bricker EM. Symposium on clinical surgery. Bladder substitution after pelvic evisceration. Surg Clin North Am 1950; 30:1511-1521.
2. Skinner DG, Studer UE, Okada K et al. Which patients are suitable for continent diversion or bladder substitute following cystectomy or other definitive local treatment? Int J Urol 1995; 2:105-112.
3. Borirakchanyavat S, Aboseif SR, Carroll PR et al. Continence mechanism of the isolated female urethra: An anatomical study of the intrapelvic somatic nerves. J Urol 1997; 158:822-826.

4. Butcher HR Jr, Sugg WL, McAffee CA et al. Ileal conduit method of ureteral urinary diversion. Ann Surg 1962; 156:682.

5. Rowland RG, Mitchell ME, Bihrle R. The cecoileal continent urinary reservoir. World J Urol 1985; 3:185-190.

6. Ahlering TE, Weinberg AC, Razor B. Modified Indiana pouch. J Urol 1991; 145:1156-1158.

7. Studer UE, Danuser H, Thalmann GN et al. Antireflux nipples or afferent tubular segments in 70 patients with ileal low pressure bladder substitutes: Long-term results of a prospective randomized trial. J Urol 1996; 156:1913-1917.

8. Grossfeld GD, Stein JP, Bennett CJ et al. Lower urinary tract reconstruction in the female using the Kock ileal reservoir with bilateral ureteroileal urethrostomy: Update of continence results and flurourodynamic findings. Urology 1996; 48:383-388.

9. Hautmann RE, Miller K, Steiner U et al. The ileal neobladder: Six years of experience with more than 200 patients. J Urol 1993; 150:40-45.

7

CHAPTER 8

Adrenal Tumors

Karl R. Herwig

Introduction

Above each kidney resides the golden yellow adrenal glands made up of an outer cortex and an inner medulla. The glands have a diverse arterial blood supply from the renal artery, aorta, and the inferior phrenic artery. The venous drainage is more constant, the right adrenal vein drains into the vena cava and the left adrenal vein drains into the left renal vein. This relationship is important diagnostically when doing venography, and surgically when dissecting the gland for removal.

Anatomy and Physiology

The adrenal cortex is arranged into three layers, the zona glomerulosa, the zona follicularis, and the zona reticularis. These cells secrete the steroid hormones, cortisol, aldosterone, and dehydroepiandrosterone, a weak androgen. Cortisol is a vital hormone necessary for intracellular metabolism; without it cells cease to function. The production of cortisol is circadian under the influence of ACTH produced by the pituitary gland. ACTH stimulates the production of cortisol, which, in turn, has the negative effect of suppressing ACTH production. This feedback mechanism occupies a fundamental concept of adrenal-cortical function and forms the basis of laboratory analysis of adrenal cortical function.

Aldosterone is the main mineral corticoid and is produced in the zona glomerulosa of the cortex. This zone contains the enzymes 18-hydroxylase and 18-hydroxydehydrogenase, which are necessary for the conversion of corticosterol to aldosterone. The enzymes do not occur in the other zones of the adrenal cortex. Aldosterone causes absorption of sodium and secretion of potassium in the distal loop of Henley of the renal tubule and maintains body fluid volume. Under ordinary conditions aldosterone secretion depends on the renin-angiotensin system. Renin is produced by the juxtaglomerular cells of the kidney in response to sodium concentration in the loop of Henley and circulating blood volume of the kidney. When activated by low sodium or low blood volume, renin activates angiotensin, causing it to form angiotensin I which is rapidly converted to angiotensin II. The angiotensin II stimulates the zona glomerulosa to secrete aldosterone and facilitates the absorption of sodium and secretion of potassium. Conversely, high sodium concentration in the loop of Henley depresses renin production.

The adrenal androgens are of no physiological purpose, but become important in pathologic conditions.

The adrenal medulla is composed of chromaffin cells, derived from neurectoderm. These cells manufacture and secrete catecholamine, epinephrine, and norepinephrine in response to stimuli, and are responsible for our ability respond to noxious events.

Urological Oncology, edited by Daniel Nachtsheim. ©2005 Landes Bioscience.

All the substances produced by the cortex and the medulla, as well as their metabolic products, are measurable in the laboratory and form the basis for identifying any abnormalities of adrenal function.

Tumors of the adrenal gland may be functional or nonfunctional, benign or malignant, primary or secondary, and can be familial with genetic predisposition. Among the functioning tumors are Cushing's syndrome, Conn's syndrome, and pheochromocytoma. Pheochromocytomas are sometimes familial and associate with other tumors of the endocrine system. Nonfunctioning tumors are often found incidentally when performing radiology studies for another condition; these are referred to as incidentalomas. Secondary tumors are malignant and most often arise from the lung or kidney.

Tumors of the Adrenal Cortex

Excess production of glucocorticoid leads to a distinct clinical picture. Among its features are:

- Muscle weakness
- Glucose intolerance
- Truncal obesity
- "Buffalo" hump
- Osteoporosis
- Renal calculi
- Mental changes
- Hirsutism
- Easy bruising
- Thin skin
- Abdominal striae
- Sexual dysfunction

This clinical picture is referred to as Cushing's syndrome. Of course, not all of the symptoms are present in the same patient nor are they present at the same time. The excess production of cortisol may be due to excess ACTH secretion from the pituitary gland, Cushing's disease, from an ectopic source such as a lung tumor, or from a tumor of the adrenal cortex itself. The vast majority of Cushing's is due to pituitary gland disease, usually a small adenoma of the anterior pituitary gland.

Another cause of Cushing's is secondary from exogenosis steroids. About 20% of hyperadrenal corticalism is due to a primary tumor of the adrenal cortex.

Adrenal tumors are independent of ACTH and are differentiated from other causes by the measurement of ACTH and manipulation of the adrenal-pituitary axis with exogenous steroid and measuring cortisol in the serum and urine. Tumors have a low ACTH level and a high urinary free cortisol level with no evidence of suppression by exogenosis steroid.

Once the diagnosis is entertained, various radiologic studies are used to identify and localize the tumor. Currently, computerized tomography (CT) scanning (Fig. 8.1) and nuclear magnetic resonance imaging (MRI) are the most useful. The isotope N59, iodocholesterol, is useful in showing aldosterone tumors (Fig. 8.2). After the presence of a tumor is confirmed, surgical removal is the proper treatment.

Excessive aldosterone production results in Conn's syndrome, characterized by hypertension, hypokalemia, and a tumor of the adrenal cortex. Not all patients with

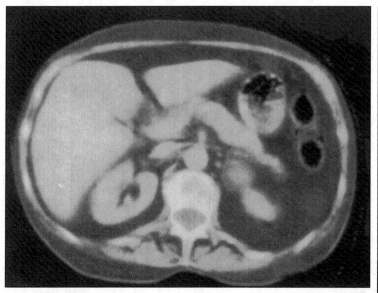

Fig. 8.1. Adrenal tumor above the left kidney, detected by C-T scan.

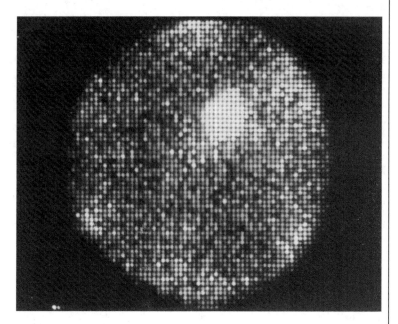

Fig. 8.2. Iodocholesterol scan of an active aldosterone producing tumor.

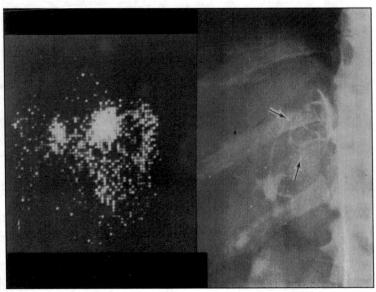

Fig. 8.3. NP59, iodocholesterol, scanning (left) and adrenal venography (right) to demonstrate an adrenal tumor.

Conn's syndrome have a tumor of the adrenal and only have hyperplasia of the zona glomerulosa. These patients should be identified before any therapy since surgery is not the treatment of hyperplasia. The patient with Conn's syndrome has markedly depressed plasma renin activity and elevated serum and urine aldosterone. These findings can be further shown by measuring renin when upright to stimulate renin production, using furosemide to lower sodium concentration, and thus stimulate renin production and captopril testing to see if aldosterone can be suppressed.

Since these laboratory studies alone will not differentiate hyperplasia from tumor, one turns to x-ray and isotope studies. The CT scan can usually demonstrate an adrenal tumor >4-5 cm. With improved computing, smaller tumors should be visible. Scanning with NP59, iodocholesterol, and dexamethasone suppression with highlight the aldosteronoma in the vast majority of patients. The best test for the presence of a tumor is adrenal venography (Fig. 8.3). However, this is technically difficult to perform and the right adrenal vein cannot always be found.

If a tumor is present and removed, hypertension is cured in most of the patients. Those not cured often have underlying essential hypertension although one does not have to contend with hypokalemia.

Adrenal cortical tumors that produce primarily androgen are very rare (Fig. 8.4). They cause masculinization and hirsutism in the female, and in some males feminization results because the excess androgen is converted to estrogen by the liver. Measurements of 17-keto steroids will tell you the amount of androgen being produced.

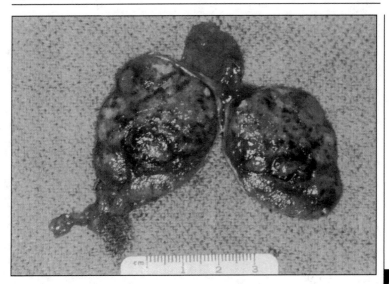

Fig. 8.4. Adrenal cortilcal tumors that produce primarily androgen are rare.

8

Tumors of the Adrenal Medulla

Pheochromocytoma is the tumor of the adrenal medulla. It produces high levels of catecholamines (epinephrine and norepinephrine) and leads to hypertension, often episodic, but usually sustained. The most common symptoms are hypertension, palpitation, severe pounding headaches, and excessive and inappropriate sweating. Familial pheochromocytomas are seen in patients with multiple endocrine neoplasia (MEN) syndrome. Usually these patients also have medullary carcinoma of the thyroid and hyperparathyroidism, MEN2A. Expression of one part of the syndrome may occur many years after another part of the syndrome, requiring lifelong follow-up for these patients. Pheochromocytomas are also seen in von Hippel-Lindau disease, von Recklinghausen's disease, and Sturge-Weber syndrome.

The biochemical diagnosis is made by measuring catecholamines, epinephrine, norepinephrine, and metanephrines in the serum and urine. Once diagnosis is made, localization of the tumor can be done with x-ray or isotope studies. Again, a CT scan will identify most of these tumors. The MRI is an excellent tool because the pheochromocytomas light up in the T2-weighted images (Fig. 8.5). MIGB, a radioisotope that is taken up by chromaffin cells is very useful in localizing pheochromocytoma, especially when the tumor is ectopic or metastatic.

The treatment of pheochromocytoma is removal of the tumor wherever it may be. The rule of tens applies to these tumors: 10% are bilateral, 10% are ectopic, and 10% are malignant. This makes localization very important preoperatively. Also, the patient with a pheochromocytoma is volume-depleted from the marked vasoconstriction. Some believe preoperative volume expansion is indicated. Others try to reduce the effect of catecholamine peripherally by using alpha-blocking agents

8

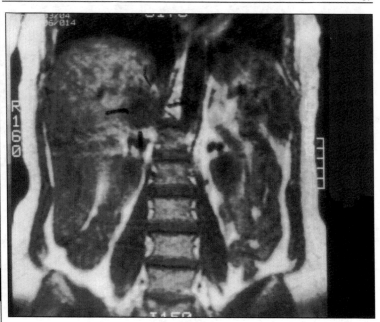

Fig. 8.5. MRI demonstrating pheochromocytoma.

such as prazosin or terazosin to allow for preoperative volume expansion. Still others are content to only control the hypertension preoperatively with calcium channel blockers. Beta blockade for cardiac arrhythmias is usually unnecessary.

Successful operation depends on good anesthesia. During the operation, liberal fluid replacement and good control of blood pressure are paramount. Nitroprusside is an excellent drug to control hypertensive episodes during the operation and reduces the need for blood pressure support post-surgery.

Incidentalomas

The increased use of scanning techniques (MRI and CT) for evaluation of abdominal and chest problems has identified many asymptomatic tumors in the adrenal glands. These tumors are referred to as incidentalomas because they are unexpected. They are found in about 2% of all abdominal scans. How to approach these tumors diagnostically and what to do with them is sometimes unclear. The majority of these tumors are small (<4 cm), and they have no evidence of function. When discovered, however, an attempt to determine their function seems prudent. Appropriate studies should include studies for Cushing's syndrome and studies for pheochromocytoma. If hypokalemia is present, aldosterone measurements are indicated. It is interesting to note that the most common functioning tumor among incidentalomas is the pheochromocytoma. Even without hypokalemia, large tumors should be removed. Most consider any tumor >6 cm in this category. The dilemma is to determine which small tumors can be safely ignored and which ones require surveillance. Certainly not all small tumors that you are going to follow conservatively require needle aspiration.

Malignant Tumors

Approximately 20% of adrenal tumors are malignant. This makes them very rare. The criteria for malignancy are not well-defined because the histologic examination of the tumor often gives no indication of malignancy. Also, some tumors, especially pheochromocytomas, have a long latent period before recurrence. The most obvious signs of malignancy are metastasis at the time of diagnosis, evidence of metastasis at the time of operation, and very large tumors. Tumors >8 cm should be considered malignant. The problem is to identify those smaller tumors that are malignant.

The prognosis for adrenal cortical or adrenal medullar tumors is poor. Wide surgical excision when initially treated is the best treatment. Chemotherapy and radiation therapies are not effective. A derivative of DDT, mitotane is toxic to adrenal cortical cells and destroys them. While not a cure, its use can reduce some of the hormonal effects of the tumor. However, it does not appear to prolong survival. Malignant pheochromocytoma symptoms can be somewhat alleviated with metrozine. Since many of these malignancies are slow growing and only locally invasive, repeat surgical removal can produce long-term survival.

Adrenal Surgery

With good localization and sound understanding of the pathophysiology of the adrenal, surgery for adrenal gland diseases is safe and can be done with a minimum of patient discomfort. Patients with Cushing's syndrome require no special preoperative preparation. The patient with Conn's syndrome should have the serum potassium corrected as well as possible. Simply giving exogenous potassium is inadequate; spironolactone must be given to retain potassium by the kidney. The preoperative preparation for pheochromocytomas is given above. There are many surgical approaches to the adrenal gland:

- Transabdominal
- Flank
- Posterior lumbar
- Thoracoabdominal
- Laparoscopic

The transabdominal approach is rarely used unless there is ectopic tumor, such as a pheochromocytoma of the organ of Zuckerkandl, or there are known metastases to be removed. The thoracoabdominal approach is excellent for large tumors, especially malignant ones. The flank approach gives good, but limited, exposure of the adrenal, but is not often used today. The posterior approach is excellent for small tumors such as aldosteronomas. There is little morbidity and the length of hospital stay is short. The development and mastery of laparoscopy has made this the treatment of choice for adrenal tumors except for the large malignant ones. In experienced hands it is safe, timesaving, especially reducing hospital time, and it returns the patient to usual activity rapidly. It is currently the treatment of choice where it is available.

The results of removal of benign adrenal tumors are excellent. Almost all are cured of the underlying disease. Even 80% of patients with hyperaldosteronism are cured of their hypertension with most losing the hypokalemia. Patients treated for pheochromocytoma are at risk of recurrence and require prolonged surveillance.

Selected Readings

1. Endocrine Web.com. 2000. This is a web site for the public but contains a wealth of information about the adrenal gland.
2. Ganguly A. Current concepts: Primary aldosteronism. N Engl J Med 1998; 339:1828-1834. This is an excellent review.
3. Goldfarb DA. Contemporary evaluation and management of Cushing's syndrome. World J Urol 1999; 17:22-25.
4. Pommier RF, Brennen MF. Management of adrenal neoplasms. Curr Probl Surg 1991; 659-734. This very thorough review has the most information about adrenal tumors in one place.
5. Staren ED, Prinz RA. Selection of patients with adrenal incidentalomas for operation. Surg Clin North Am 1995; 75:499
6. Ulchaker JC, Goldfarb DA, Bravo EL, Novick AC. Successful outcomes in pheochromocytoma surgery in the modern era. J Urol 1999; 161:764-767.
7. Zografos GC, Driscoll DL, Karakousis CP, Huben RP. Adrenal adenocarcinoma. A review of 53 cases. J Surg Oncol 1994; 55:160-164.

8

CHAPTER 9

Penile and Urethral Cancer

Daniel J. Cosgrove and Joseph D. Schmidt

Introduction

Penile and urethral cancers are extremely rare malignancies of the genitourinary tract. Due to the paucity of cases, a standardized approach to management of these malignancies is still under debate and continues to evolve. Squamous cell carcinoma is the most common cancer in both the penis and urethra.

Penile Cancer

Epidemiology

Cancer of the penis accounts for 2% of all genitourinary malignancies. In Europe and North America, penile cancer accounts for approximately 0.4-0.6% of all malignancies. Among males in the United States, the incidence is 1 to 2 cases per 100,000 per year. It is more prevalent in less developed countries. Penile cancer is the most common malignancy in the adult African male and accounts for 22% of male cancers in China and 12% among Indian Hindus.

Some studies have shown a preponderance of blacks afflicted with this disease compared with whites. This may be a reflection of socioeconomic differences rather than a racial predisposition.

Age at Presentation

Penile cancer is a disease of older men, almost exclusively affecting those over the age of 60. In India, however, men between the ages of 45 and 55 are most commonly affected, and, in Russia, one-third of the patients are less than 40 years old. The disease has been reported rarely in children.

Etiology

The vast majority of penile cancers occur in uncircumcised men, implicating smegma and poor hygiene as causative factors. Among men who were circumcised in the neonatal period, penile carcinoma is extremely rare. Circumcision performed at puberty or in adulthood offers little or no protection from development of the disease. The virtual absence of the disease in Jews is a reflection of the ritual circumcision performed on the eighth postnatal day. Moslems who become circumcised between the ages of 4 and 12 have higher incidences of penile cancer than do Jewish males. Similarly, the extremely low incidence of this disease in North America reflects the fact that the majority of these male children are circumcised in the neonatal period.

Urological Oncology, edited by Daniel Nachtsheim. ©2005 Landes Bioscience.

Phimosis, poor hygiene, and retained smegma have all been implicated in the etiology of penile cancer. It is more likely that they each play some role in carcinogenesis resulting from chronic inflammation and irritation that occurs with retained secretions and bacteria beneath the foreskin (prepuce).

Although smegma has been found to be carcinogenic in animal models, to date no component of human smegma responsible for malignant transformation has been identified. Although a history of trauma has been suggested as a causative factor, this is more likely coincidental,

There is no concrete evidence that penile cancer is associated with sexually transmitted diseases such as syphilis, granuloma inguinale, or chancroid. Some studies have shown a relationship between herpesvirus and penile cancer, but others have failed to demonstrate this. There is a known correlation between human papillomavirus (HPV) and squamous cell carcinoma of the cervix, and recent studies have been able to demonstrate HPV DNA sequences in patients with penile cancer. Furthermore, wives of patients with penile cancer have been noted to be at increased risk of developing cancer of the cervix. Thus, evidence is accumulating that sexual transmission of HPV may play a role in the etiology of penile cancer.

Pathology

More than 95% of penile cancers are squamous cell, with the characteristic histology of keratinization, epithelial pearl formation, loss of cellular polarity, abnormal mitotic activity and disruption of the normal rete pegs. Most penile carcinomas are of low grade. As of yet, a relationship between tumor grade (on the basis of histologic differentiation) and survival has not been shown.

Other tumors found on the penis include melanomas, basal cell carcinomas, Bowen's disease (carcinoma in situ), mesenchymal tumors (including leiomyoma sarcomas, fibrosarcomas and Kaposi's sarcoma), metastatic lesions and leukemic or lymphoma infiltrates.

Several histologically benign penile lesions have been shown to have malignant potential or close association with squamous cell carcinoma. These include the penile cutaneous horn, leukoplakia, erythroplasia of Qeyrat, Bowen's disease, balanitis xerotica obliterans, giant condyloma acuminatum (Buschke-Lowenstein tumor), and verrucous carcinoma. Management of these lesions for the most part involves local excision with or without adjuvant topical medications and close follow-up.

Clinical Presentation

The presence of a penile lesion itself usually brings the patient to medical attention. On gross appearance, the tumor may be nodular, ulcerative or fungating. The tumors may appear anywhere on the penis, but are found most commonly on the glans or inner surface of the prepuce. Less commonly, tumors are found on the coronal sulcus and penile shaft. Other symptoms at presentation include pain, phimosis, bleeding, a malodorous discharge or difficulty with urination. Symptoms related to inguinal lymphadenopathy are less common presenting manifestations.

The disease is often locally advanced at the time of presentation, reflecting the fact that a large majority of patients delay seeking medical attention. Explanations for this delay include personal neglect, embarrassment, fear, guilt, and ignorance (Fig. 9.1A,B).

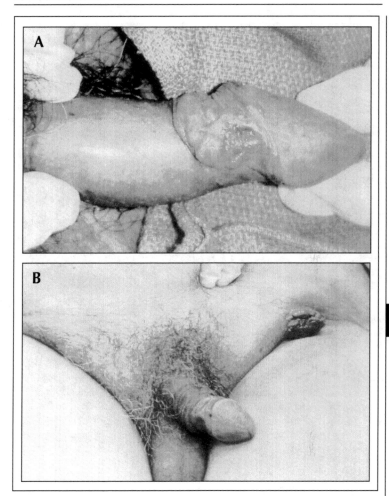

9

Fig. 9.1. A 70-year-old man with a 1-year history of a penile lesion and 1-month history of left groin mass. Pathology following circumcision revealed squamous cell carcinoma of the penis (A). The left groin mass was also positive for squamous cell carcinoma (B).

Diagnosis

Diagnosis is confirmed with incisional or excisional biopsy. Microscopic examination of the biopsy and determination of depth of invasion are mandatory before the initiation of any therapy. A thorough physical exam will help determine the extent of local invasion as well as the status of inguinal lymphadenopathy. Computerized tomography (CT) (Fig. 9.2) and magnetic resonance imaging (MRI) are used to further evaluate local spread as well as the pelvic and abdominal lymph nodes.

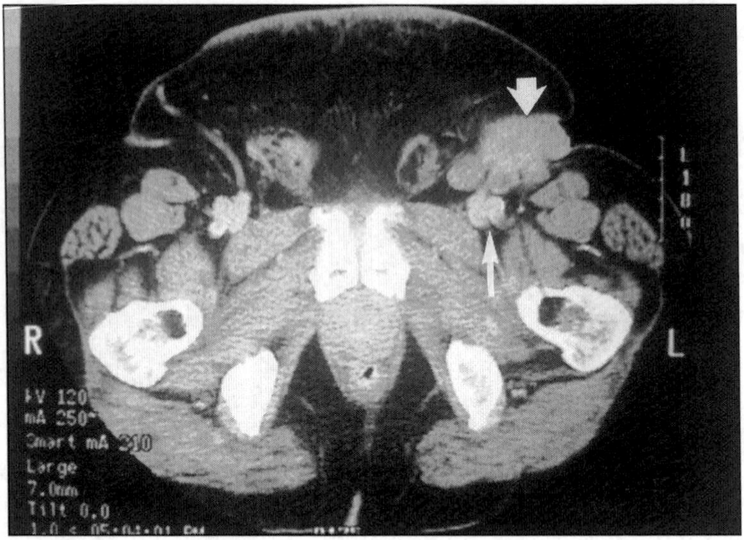

Fig. 9.2. CT of the pelvis of the same patient as in Figure 9.1 revealing a 3 x 6 x 6 cm soft tissue mass anterior and lateral to the left iliac vessels. The mass appears to arise from and encase several lymph nodes.

Chest radiography and CT are also used to determine the presence of metastatic disease. CT-guided fine needle aspiration of groin nodes may obviate the need for groin dissection or at least modify the surgical plan. Lymphangiography is rarely performed today. Initial laboratory studies include a hematological profile, serum chemistries and liver function tests. In patients with penile cancer, these studies are usually normal. However, hypercalcemia without bony metastases has been reported in association with penile cancer.

Staging

There is currently no universal staging system for penile cancer, although the two most commonly applied are the Jackson and UICC/TNM staging systems (Tables 9.1 and 9.2).

Natural History

Squamous cell carcinoma of the penis grows slowly. Deep penetration of the tumor is delayed by Buck's fascia, a tough membranous sheath surrounding the corpora cavernosa. Penetration of Buck's fascia and the tunica albuginea allows for invasion into the very vascular corpora and the potential for hematogenous spread. Disease spread beyond the primary site is usually via lymphatic channels rather than hematogenous. Lymphatics from the prepuce and the skin of the shaft drain into the superficial inguinal nodes. Drainage from the glans joins the lymphatics from the corporal bodies and drains by way of the superficial inguinal nodes into the deep inguinal nodes and to the pelvic nodes. There is no "skip" drainage, but multiple collateral channels exist, causing bilateral inguinal drainage.

Table 9.1. TNM staging of carcinoma of the penis

Primary Tumor (T)

TX	Primary tumor cannot be assessed
T0	No evidence of primary tumor
Tis	Carcinoma in situ
Ta	Noninvasive verrucous carcinoma
T1	Tumor invades subepithelial connective tissue
T2	Tumor invades corpus spongiosum or cavernosum
T3	Tumor invades urethra or prostate
T4	Tumor invades other adjacent structures

Regional Lymph Nodes (N)

NX	Regional lymph nodes cannot be assessed
N0	No regional lymph node metastasis
N1	Metastasis in a single superficial, inguinal lymph node
N2	Metastasis in multiple or bilateral superficial inguinal lymph nodes
N3	Metastasis in deep inguinal or pelvic lymph nodes(s) unilateral or bilateral

Distant Metastasis (M)

MX	Distant metastasis cannot be assessed
M0	No distant metastasis
M1	Distant metastasis

Stage Grouping

Stage 0	Tis	N0	M0
	Ta	N0	M0
Stage I	T1	N0	M0
Stage II	T1	N1	M0
	T2	N0	M0
	T2	N1	M0
Stage III	T1	N2	M0
	T2	N2	M0
	T3	N0	M0
	T3	N1	M0
	T3	N2	M0
Stage IV	T4	Any N	M0
	Any T	N3	M0
	Any T	Any N	M1

9

Used with permission from the American Joint Committee on Cancer (AJCC®), Chicago, IL. From AJCC® Cancer Staging Manual, 5th ed. Philadelphia: Lippincott-Raven Publishers, 1997:215-216.

Table 9.2. Jackson staging of carcinoma of the penis

Stage I
Tumor limited to glans penis and/or prepuce.
Stage II
Invasion into shaft or corpora. Negative nodes.
Stage III
Tumor confined to penis with proven regional nodal disease.
Stage IV
Invasion from shaft with inoperable regional modal disease or distant metastases.

Metastatic enlargement of regional nodes may eventually lead to sepsis and hemorrhage from erosion into the femoral vessels. The average life expectancy of a patient with nodal metastases is two years.

At least half of all patients with penile carcinoma will present with inguinal lymphadenopathy. This may be due to infection and inflammation rather than metastatic spread. Distant metastases to liver, lung, brain or bone are extremely rare, occurring in 1-10% of reported cases.

Treatment

Treatment is dependent on both the extent of the primary lesion and the nodal involvement. Surgery, radiotherapy and chemotherapy have all been used in various combinations in the treatment of penile carcinoma. Due to the paucity of cases, however, there are no randomized, prospective trials comparing these modalities.

Management of the Primary Lesion

Partial or total penectomy remains the gold standard for treatment of penile cancer. Due to the severe disfigurement associated with penile amputation, organ-sparing techniques are being investigated, including Mohs' micrographic surgery (MMS) and non-surgical techniques such as x-ray, laser, and cryotherapy.

Surgery

Circumcision

In a few select cases with tumor involving only the prepuce, complete excision has been accomplished with circumcision alone. Careful examination of the margins of resection is imperative, and more radical therapy is mandatory if the margins are not clear of tumor. Post-circumcision recurrence rates of 32-50% have been reported. Local wedge resection should be avoided, as recurrence rates of up to 50% have been reported.

Partial Penectomy

For stage I and II tumors involving the glans or distal shaft, partial penectomy with a 2 cm proximal margin is the surgical treatment of choice. Frozen section of the proximal margin is necessary to confirm adequate tumor resection. Adhering to this guideline minimizes tumor recurrence at the line of resection, even with deep corporal invasion, and also leaves an adequate penile stump for sexual function and upright urination. In the absence of positive inguinal nodes, 70-80% five-year survival rates following partial penectomy have been reported.

Radical (Total) Penectomy

For primary lesions extending to the base of the penis, tumors in which 2 cm margins are not possible or tumors that preclude the formation of a penile stump adequate for erect micturition, radical penectomy (and perineal urethrostomy) is the procedure of choice. During radical dissection, the testes may require removal en bloc with the penis, as their preservation frequently means inadequate removal of the scrotal skin encroaching on the tumor margins. In addition, the retained scrotum can hang in front of the newly created perineal urethrostomy and can be excoriated with constant exposure to urine. Partial and total penectomy are psychologically traumatic events. Adequate counseling and support services should be made available to all patients undergoing such procedures.

Mohs' Micrographic Surgery (MMS)

MMS is a method of removing skin cancers by excising tissues in thin layers. It is an attractive modality for the treatment of some small, select penile lesions. It can excise extensions of the primary lesion with cure rates equivalent to more radical surgery, while at the same time providing improved functional shaft length and cosmesis compared to the more radical treatments. Mohs' reported a 100% cure rate in lesions <1 cm but only 50% in lesions >3 cm. Furthermore, in patients who undergo this procedure for the larger lesions, penile reconstruction is often difficult. MMS is ideally suited for small, distally located carcinomas.

Radiation Therapy

Like MMS, primary radiation therapy can allow preservation of penile structure and function in a select group of patients. Various techniques of administering radiation therapy to the penis have been reported, including external beam, electron beam, iridium mold and interstitial therapy. Because squamous cell carcinoma is characteristically a radioresistant tumor, the dosage of radiotherapy required to treat the tumor (i.e., 6,000 rad) may cause severe complications including urethral fistulas, strictures or stenosis as well as penile necrosis, edema and pain. Secondary penectomy following complications associated with radiation therapy has been reported. Another drawback of radiation therapy is in the difficulty of distinguishing a post-irradiation scar or ulcer from tumor recurrence, often requiring repeat biopsy. The patients for whom radiation therapy should be considered are:

1. the young individual with a small (2-3 cm) superficial, non-invasive lesion on the glans or coronal sulcus,
2. the patient who refuses surgery as the initial form of therapy, and
3. the patient with an inoperable tumor or distant metastases who requires local therapy but has a desire to retain his penis.

Laser Surgery

Lasers are currently being employed to treat benign and pre-malignant penile lesions as well as stage Tis, Ta, T1 and some T2 penile cancers. Like MMS and radiation therapy, lasers also have the potential advantage of destroying the primary lesion while preserving normal penile structure and function. Four different types of lasers are currently being used to treat penile lesions. These include CO_2, Nd:YAG, argon and potassium titanyl phosphate (KTP) lasers. It is essential to adequately stage the depth of tumor penetration with deep biopsy before using laser therapy, as the depth of laser destruction may be difficult to determine.

Management of Regional Lymph Nodes

The presence of inguinal metastases portends a poor prognosis for the patient. Patients with penile carcinoma and inguinal metastases who choose not to undergo lymphadenectomy rarely survive two years and almost never survive five years. Those patients with clinically palpable adenopathy and histologically proven inguinal metastases who do undergo lymphadenectomy achieve a 20-50% five-year survival rate. Lymphadenectomy may be curative in some cases of penile carcinoma and should be recommended whenever possible. Fifty percent of patients with penile carcinoma present with inguinal lymphadenopathy. In half of these cases, inguinal lymphadenopathy represents inflammation associated with an infected penile lesion. As such, reevaluation of the inguinal nodes should be performed 4-6 weeks following treatment of the primary lesion and a course of antibiotics, allowing time for the infection and inflammation to subside. Persistent adenopathy following treatment of the primary lesion and the 4- to 6-week course of antibiotics is most often due to metastatic disease and warrants biopsy and therapy.

Inguinal lymphadenectomy can range from biopsy of the sentinel node to a radical dissection. As the procedure itself can cause significant morbidity, the choice of procedure should be tailored to the individual patient. Superficial inguinal lymphadenectomy entails excision of the nodes superficial to the deep fascia. In deep inguinal node dissection, the nodes below the deep fascia are also excised. In iliac or pelvic lymphadenectomy, the external iliac and obturator nodes are removed. Radical inguinal lymphadenectomy includes superficial and deep lymphadenectomies plus pelvic lymphadenectomy. Ilioinguinal lymphadenectomy has the potential for significant morbidity including phlebitis, pulmonary embolism, wound infection, flap necrosis and chronic lymphedema. Recent improvements and modifications in surgical technique have significantly reduced the complication rate. These improvements include preservation of the dermis, Scarpa's fascia, and the saphenous vein.

The selection of patients who will potentially benefit from lymphadenectomy has been a topic of much debate. The current recommendation is to select a treatment plan based on clinical stage and tumor grade.

Tis, Ta, T1: N0M0 (Jackson Stage I, II)

For these patients with no palpable lymphadenopathy and only superficial disease, immediate inguinal lymphadenectomy is not warranted. Because nodal metastases may eventually occur, patients in this group should be taught careful self-examination of the inguinal areas to monitor for metastases and should be followed at two- to three-month intervals.

T2, T3 N0M0 (Jackson Stage I, II)

These lesions are limited to the glans or shaft with no clinically palpable adenopathy. Pathologically they show invasion of corpus spongiosum, cavernosum, urethra or prostate. Studies have shown nodal metastases in up to two-thirds of these patients with clinically non-palpable nodes. For this reason, some authors believe that immediate bilateral adjunctive lymph node dissection is warranted. However, the currently available clinical data have not yet shown that early "prophylactic" lymphadenectomy provides any significant survival advantage over delayed lymphadenectomy (after lymph glands have become clinically suspicious).

Stage Tis, Ta, T1-3, N1-3, M0 (Jackson Stage III)

For those patients with clinically palpable bilateral lymphadenopathy after an appropriate course of antibiotics, bilateral radical lymph node dissections are warranted. If lymphadenopathy is clinically unilateral, then following an ipsilateral radical node dissection with pathologically negative nodes, a contralateral superficial lymph node dissection or deep node dissection with limited boundaries can be considered.

Any T, Any N, M1; T4; Inoperable N (Jackson Stage IV)

For patients with distant metastases, inoperable inguinal adenopathy due to fixation or invasion, and extensive adjacent organ invasion, treatment is often limited to palliative chemotherapy or radiotherapy. Aggressive combined modality therapy is, however, performed especially for the younger patient. This involves neoadjuvant chemotherapy followed by inguinal and pelvic lymphadenectomy with consideration of hemipelvectomy or hemicorporectomy.

Radiation Therapy

Primary radiation therapy as treatment for inguinal lymphadenopathy is not an accepted therapeutic modality. Concerns related to the use of primary radiation therapy include the inaccuracy of clinical staging, the ineffectiveness of radiation therapy in the setting of infected lymph nodes and the skin maceration and ulceration that often develops from radiation especially to the groin areas. Furthermore primary radiation therapy of the inguinal areas has been shown to be less effective therapeutically than lymph node dissection alone. In the situation of inoperable inguinal nodes, radiation therapy can be used for palliation.

Chemotherapy

Information related to the management of penile carcinoma with chemotherapeutic agents is sparse due to the rarity of the disease. Bleomycin, methotrexate and cis-platinum have been shown to be active against other squamous cell carcinomas and have thus been employed for the treatment of metastatic penile carcinoma. As single agents, partial responses have been reported in the range of 15-27% for cisplatin, 0-62% for methotrexate and 20-21% for bleomycin. Complete remissions are rare and toxicities are usually high. Current investigations involve combinations of these three agents.

Urethral Cancer

Introduction

Like penile carcinoma, urethral carcinoma is extremely rare. It has the distinction of being the only urologic malignancy more common in females than in males. As with penile carcinoma, squamous cell carcinoma is the most common cancer in the urethra. Urethral cancer in both males and females tends to invade locally and metastasize to regional lymph nodes early in the course of the disease. At the time of diagnosis, most of these tumors are far advanced locally. Despite aggressive management, urethral cancer generally caries a poor prognosis.

Carcinoma of the Male Urethra

Epidemiology

Urethral cancer accounts for less than one percent of cancers in males. To date, less than 600 cases have been reported. It is a disease of the elderly with most reported cases in men over 60 years of age although cases have been reported in boys as young as 13 and men as old as 90. There is no apparent racial predisposition.

Etiology

The exact cause of urethral cancer is unknown; however, there has been shown to be an association between chronic infection and inflammation and urethral cancer. Many patients with urethral cancer have a prior history of sexually transmitted disease, urethritis or urethral stricture. Urethral strictures are commonly seen in men with urethral cancer with the incidence ranging from 24-88%. It should be noted that the bulbomembranous urethra, which is the most common site of stricture formation, is also the most common site of urethral cancer.

Anatomy and Pathology

The male urethra can be divided anatomically into anterior and posterior portions (Fig. 9.3). The anterior urethra comprises the glanular, pendulous and bulbous segments, while the posterior urethra contains the membranous and prostatic segments. The mucosal cell type varies along the urethra; the meatus is lined with stratified squamous epithelium, while the penile, bulbous and membranous urethra are lined with pseudostratified or stratified columnar epithelium. The prostatic urethra is lined with transitional epithelium.

Histologically, 80% of male urethral cancers are squamous cell carcinoma. It is most frequently located in the bulbomembranous region of the urethra suggesting that there has been a metaplastic differentiation from the existing pseudostratified or stratified columnar epithelium. Fifteen percent of urethral cancers are transitional cell carcinoma and usually arise in the prostatic urethra. It is important to distinguish urethral transitional cell carcinoma from transitional cell carcinoma arising from the bladder as the prognosis for each differs. Five percent of urethral cancers are adenocarcinoma of uncertain origin and undifferentiated tumors.

Overall, 60% of male urethral cancers are found in the bulbomembranous region, 30% in the penile urethra and 10% in the prostatic urethra.

Male urethral cancers tend to remain localized until late in the disease process. Initial spread is by direct extension into adjacent tissues. Carcinoma of the urethra can extend into the periurethral tissues as well as to the urogenital diaphragm, prostate, perineum, and scrotal skin. Lymphatic spread occurs late in the disease. In general, anterior urethral tumors drain into the inguinal lymph nodes, whereas posterior tumors drain into the iliac and pelvic lymph nodes. The perimeatal lymphatics drain into the superficial inguinal and subinguinal glands while the glans and corpora drain into the deep inguinal nodes below the fascia lata. Both of these groups of nodes then drain into the external iliac nodes. The lymphatics from the bulbous, membranous and prostatic urethra drain in parallel to the dorsal vein of the penis and lead to the external iliac nodes, while collateral lymphatics pass via the pudendal vessels to the obturator, internal iliac and presacral glands.

Hematogenous spread with disseminated disease is uncommon but is seen more frequently with adenocarcinoma of the urethra. When it does occur, the most common sites of spread are the lung, liver, kidney, adrenals and pleura.

Clinical Presentation

There are no pathognomonic signs or symptoms of urethral cancer. The most common presenting signs and symptoms are from irritation and urethral

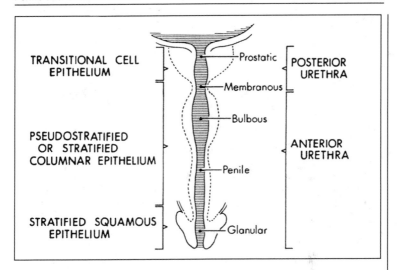

Fig. 9.3. Anatomy of the male urethra showing histology of the mucosa and anatomic divisions. From: Cancer 1980; 45(7):1966, with permission.

obstruction as well as bleeding, hematospermia or discharge. Less frequently patients may present with recurrent infections or a periurethral abscess. It is important to remember that many of the early signs and symptoms of urethral cancer are the same as those of more benign urethral disease such as stricture and urethritis. The clinician must have a high index of suspicion when a patient presents with urethral bleeding and no prior history of urethral disease or trauma, or if there is increasing difficulty dilating a stricture.

Diagnosis

Visual inspection is only helpful in very distal urethral cancers. Physical examination must include thorough palpation of each section of the urethra together with careful examination of the inguinal nodes. Unlike penile cancer, palpable inguinal lymphadenopathy usually represents metastatic disease rather than infection. Rectal exam may be very helpful in assessing spread of tumor into adjacent tissues. A urethrogram may help delineate the extent of the tumor. Urine cytologies may be helpful in making the diagnosis, but definitive diagnosis is made with biopsy of the tumor under direct vision during cystourethroscopy.

Differential diagnosis includes stricture, periurethral abscess, calculus, tuberculosis, prostate cancer, condyloma and foreign body.

Staging

Once the diagnosis has been confirmed, various studies are performed to complete the staging of the disease. These include chest x-ray and serum chemistries, computerized tomography (CT) or magnetic resonance imaging (MRI) to evaluate for pelvic and paraaortic lymph node involvement, NM to evaluate local soft tissue and a bone scan to look for bone metastases.

Urethral cancer is staged using the TNM classification, which evaluates for depth of invasion of the primary tumor, the presence or absence of regional lymph node involvement, and distant metastases (Table 9.3).

Treatment

Treatment depends on the location and stage of the tumor. In general, anterior tumors present earlier and have a better prognosis than posterior tumors. Surgery is the mainstay of treatment for male urethral cancer.

Carcinoma of the Distal Urethra (Anterior Urethra)

Superficial papillary or in situ cancers of the distal urethra can be adequately treated with transurethral resection (TUR). However, such early presentations are quite rare. Sleeve resection for a localized lesion is also possible with either an end-to-end reanastomosis of the urethra or complete urethral reconstruction.

Urethral tumors infiltrating the corpus and localized to the distal half of the penis are best treated by partial penectomy leaving a 2-cm margin proximal to visible tumor or palpable induration.

For infiltrating tumors of the distal penile urethra or tumor involving the entire urethra, radical penectomy with perineal urethrostomy is the treatment of choice. The testes should be removed when performing radical amputation.

Ilioinguinal lymphadenectomy is indicated only for clinically suspicious nodes and is carried out 6 weeks after amputation. Prophylactic inguinal lymphadenectomy has not been shown to be of benefit in urethral cancer patients.

9

Carcinoma of the Bulbomembranous Urethra (Proximal Urethra)

Most cancers of the bulbomembranous urethra are of the squamous variety and present at higher stages than the anterior cancers, usually with positive lymph nodes at the time of diagnosis. Superficial tumors of this region may be treated with TUR or segmental resection; however, they are rare. Survival is generally poor for patients with proximal urethral cancer and as such treatment has historically been focused on palliation. More recently, radical surgery combined with radiation or chemotherapy is being used in treating these tumors.

The current recommendation is to perform the surgery in two stages. Stage I involves a staging laparotomy to evaluate the pelvic and iliac nodes to determine if further surgery is feasible. Extensive pelvic involvement or positive lymph nodes above the iliac bifurcation preclude any surgery other than palliation. This stage concludes with supravesical urinary diversion.

Following surgery, external beam radiation is administered to the tumor mass over a two-week period.

The second stage of the surgery is performed 3-4 weeks after the initial staging laparotomy. Surgery usually involves radical cystoprostatectomy, pelvic lymphadenectomy, total penectomy with excision of the scrotum as well as an ellipse of perineal skin. Resection of the pubic symphysis, anterior rami and adjacent urogenital diaphragm may improve the surgical margin of resection and control of the disease.

Chemotherapy is also being combined with radiation and surgery in an attempt to improve outcomes.

Table 9.3. TNM staging of carcinoma of the urethra

Definition of TNM
Primary Tumor (T) (Male and Female)

TX	Primary tumor cannot be assessed
T0	No evidence of primary tumor
Ta	Noninvasive papillary, polypold, or verrucous carcinoma
Tis	Carcinoma in situ
T1	Tumor invades subepithelial connective tissue
T2	Tumor invades any of the following: corpus spongiosum or cavernosum, prostate, periurethral muscle
T3	Tumor invades any of the following: corpus cavernosum, beyond prostatic capsule, anterior vagina, bladder neck
T4	Tumor invades other adjacent organs

Transitional Cell Carcinoma of the Prostate

Tis pu	Carcinoma in situ, involvement of the prostatic urethra
Tis pd	Carcinoma in situ, involvement of the prostatic ducts
T1	Tumor invades subepithelial connective tissue
T2	Tumor invades any of the following: prostatic stroma, corpus spongiosum, periurethral muscle
T3	Tumor invades any of the following: corpus cavernosum, beyond prostatic capsule, bladder neck (extraprostatic extension)
T4	Tumor invades other adjacent organs (invasion of the bladder)

Regional Lymph Nodes (N)

NX	Regional lymph nodes cannot be assessed
N0	No regional lymph node metastasis
N1	Metastasis in a single lymph node, 2 cm or less in greatest dimension
N2	Metastasis in a single node more than 2 cm in greatest dimension, or in multiple nodes

Distant Metastasis (M)

MX	Distant metastasis cannot be assessed
M0	No distant metastasis
NI	Distant metastasis

Stage Grouping

Stage 0a	Ta	N0	M0
Stage 0is	Tis	N0	M0
	Tis pu	N0	M0
	Tis pd	N0	M0
Stage I T1		N0	M0
Stage II T2		N0	M0
Stage III	T1	N1	M0
	T2	N1	M0
	T3	N0	M0
	T3	N1	M0
Stage IV	T4	N0	M0
	T4	N1	M0
	Any T	N2	M0
	Any T	Any N	M1

Used with permission from the American Joint Committee on Cancer (AJCC®), Chicago, IL. From AJCC® Cancer Staging Manual, 5th ed. Philadelphia: Lippincott-Raven Publishers, 1997:248.

9

Primary Carcinoma of the Prostatic Urethra

Diagnosis of primary prostatic urethral carcinoma is based on finding a solitary tumor, usually transitional or adenocarcinoma, in the prostatic urethra with no evidence of co-existing or pre-existing urothelial tumors. The rare superficial tumor can be managed with transurethral resection alone. More often the tumor involves the bulk of the prostate and can extend into bulbomembranous urethra or into the base of the bladder. With these more extensive tumors, cystoprostatectomy and total urethrectomy is the treatment of choice. Multimodality therapy using a combination of surgery, radiation and chemotherapy is being investigated.

Radiation Therapy

Radiation therapy has been used as primary therapy, as an adjunct to surgical excision and for palliation. Long term results of radiation therapy are mixed, however the most promising results are seen with distal lesions for which radiation therapy has had similar results to surgical excision.

Prognosis

For anterior tumors, five-year survival is approximately 50%. The overall survival for posterior urethral tumors is poor with only 8% living at five years. At least 50% of patients with posterior urethral tumors will have extensive disease amenable to palliation only at the time of diagnosis. Of these, 96% will be dead three months from diagnosis.

Carcinoma of The Female Urethra

Epidemiology

Carcinoma of the female urethra occurs 4 times more often than male urethral cancer. Urethral cancer accounts for less than 0.02% of all genital cancers in women. Female urethral cancer is seen more frequently in whites than in blacks, and, as with males, is seen more frequently in the elderly with the majority of patients being older than 50 years of age.

Etiology

No definite etiologic factors have been identified. There has been some suggestion that chronic irritation, the trauma of intercourse and childbirth, and the vulnerability to urinary tract infection due to the short female urethra may all predispose to malignancy. Urethral diverticula may be the site of malignancy resulting from secondary inflammation and stone disease.

Anatomy and Pathology

The distal two-thirds of the female urethra is lined by stratified squamous epithelium and the proximal one third is lined with transitional epithelium. The anterior female urethra is defined as the distal one third of the urethra and 50% of all female urethral cancers arise in this area (Fig. 9.4). The lymphatics from the anterior urethra and labia drain into superficial and deep inguinal nodes. The proximal, or posterior urethra, drains into the deep pelvic nodes consisting of the obturator, external and internal iliac nodes.

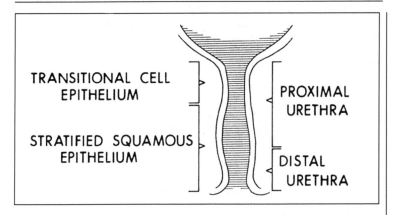

Fig. 9.4. Anatomy of the female urethra showing histology of the mucosa and anatomic divisions. From: Cancer 1980; 45(7):1970, with permission.

Sixty percent of female urethral cancers are squamous cell, 20% are transitional cell carcinoma, and 10% are adenocarcinoma. The remainder consist of melanoma, sarcomas and undifferentiated tumors. Histology does not seem to affect prognosis and management is the same regardless of histologic type.

Clinical Presentation
The signs and symptoms of female urethral cancer are variable as with male urethral cancer. Patients may present with bleeding, lower urinary tract symptoms such as dysuria, urgency or frequency and less often a palpable urethral mass or induration.

Diagnosis
On physical examination it may be difficult to distinguish between tumors of the urethra and those of the vulva or vagina. Evaluation requires a pelvic examination under anesthesia, a thorough evaluation of inguinal nodes and cystourethroscopy and biopsy of the suspected lesion.

Staging
Serum chemistries, chest x-ray and CT of the abdomen and pelvis should be obtained. Female urethral cancer is staged using the TNM staging system (Table 9.3).

Treatment
Treatment and prognosis are related to location and stage of the disease rather than histological type.

Anterior Urethral Cancers
Selected patients with distal cancers can be treated with transurethral resection or local excision alone. Tumors of the distal urethra with only limited invasion may be treated with partial urethrectomy. Up to two-thirds of the distal urethra may be

excised while still preserving continence. This may be combined with interstitial or external beam irradiation. Brachytherapy has also been suggested for small, distal urethral cancers. The incidence of lymph node involvement with distal urethral cancers is low.

Posterior Urethral Cancers

For proximal urethral cancers or those invasive into adjacent structures, more aggressive therapy including preoperative external beam radiotherapy (50-65 Gy) followed by exenterative surgery is required. Single modality therapy has been associated with poor survival. When surgery is considered, it is best performed in a staged fashion as with advanced male urethral cancer. An initial staging laparotomy and urinary diversion should be performed followed several weeks later with radical anterior exenteration removing the bladder, uterus, vagina, and anterior perineum.

Radiation Therapy

Radiation therapy alone with brachytherapy, interstitial or intracavitary irradiation has been shown to be sufficient in treating small lesions of the distal urethra. Proximal urethral tumors or higher stage tumors require combined external beam irradiation and brachytherapy. Complications from radiation therapy include bowel obstruction, fistula formation, urethral stricture, and incontinence.

Prognosis

Five-year survival for anterior urethral cancers is approximately 50%. For posterior urethral cancers, however, five-year survival is only 10%.

Combined Modality Therapy—Future Trends

Radical surgery or irradiation alone for advanced urethral cancers are associated with high morbidity and low tumor control rates. As such, investigations are ongoing to combine radiation and chemotherapy [methotrexate, vinblastine, doxorubicin and cisplatin (MVAC), mitomycin-C, gemcitabine and 5-fluorouracil] followed by surgery in an attempt to improve outcomes.

Selected Readings

1. Kini MG. Cancer of the penis in a child, aged two years. Indian Med Gaz 1944; 79:66.
2. Narasimharao KL, Chatterjee H, Veliath AJ. Penile carcinoma in the first decade of life. Br J Urol 1985; 57:358.
3. Villa LL, Lopes A. Human papillomavirus DNA sequences in penile carcinomas in Brazil. Int J Cancer 1986; 15:856-855.
4. Martinez I. Relationship of squamous cell carcinoma of the cervix uteri to squamous cell carcinoma of the penis among Puerto Rican women married to men with penile cancer. Cancer 1969; 24:777-780.
5. Rudd FV, Root RK, Skoglund RW Jr et al. Tumor induced hypercalcemia. J Urol 1972; 107:986-989.
6. Malakoff AF, Schmidt JD. Metastatic carcinoma of the penis complicated by hypercalcemia. Urology 1975; 5:510-513.
7. Narayana AS, Olney LE, Loening SA et al. Carcinoma of the penis: Analysis of 219 cases. Cancer 1982; 15:2185-2191.

9

8.   McDougal WS, Kircher FK Jr, Edwards RH et al. Treatment of carcinoma of the penis: The case for primary lymphadenectomy. J Urol 1986; 136:38-41.

9.   Jensen MO. Cancer of the penis in Denmark 1942 to 1962 (511 cases). Dan Med Bull 1977; 24:66-72.

10. Sullivan J, Grabstald H. Management of carcinoma of the urethra . In: Skinner DG, deKernion JB, eds. Genito-Urinary Cancer. Philadelphia: W.B. Saunders, 1978.

11. Herr HW, Fuks Z, Scher HI. Cancer of the urethra and penis. In: Devita VT Jr, Hellman S, Rosenberg SA, eds. Cancer, Principles and Practice of Oncology. 5th ed. Philadelphia: Lippincott, 1997.

12. Lynch DF, Schellhammer PJ. Tumors of the penis. In: Walsh PC, Retik AL, Vaughan ED, Wein AJ, eds. Campbell's Urology. 7th ed. Philadelphia: W.B. Saunders, 1998:2453-2485.

13. Sarosdy MF. Urethral carcinoma. In: Ball TP, ed. AUA Update Series. AUA Office of Education. Houston: AUA Office of Education, Volume 6, Lesson 13, 1987.

14. Schmidt JD, Blandy JP, Hope-Stone HF. Tumors of the penis and urethra. In: Moossa AR, Schmpff SD, Robson MD, eds. Comprehensive Textbook of Oncology. 2nd ed. Baltimore: Williams and Wilkins, 1991:1100-1109.

15. Thompson IM, Fair WR. Penile carcinoma. In: Ball TP, ed. AUA Update Series. Houston: AUA Office of Education, Volume 9, Lesson 1, 1990.

9

CHAPTER 10

Testicular Tumors

Daniel J. Cosgrove. and Joseph D. Schmidt

Introduction

Testicular cancers are relatively rare, but are the most common solid tumor in males aged 15-35. They have become the most curable of all urologic cancers due to effective diagnostic techniques, sensitive tumor markers, effective multidrug chemotherapy, and modifications of surgical technique. More than 95% of testicular tumors are derived from germinal tissue, the remainder originating from nongerminal or stromal cells. The testicular germ cell tumors are classified as seminomatous or non-seminomatous. Seminomas are extremely sensitive to radiation therapy, whereas non-seminomatous germ cell tumors are responsive to platinum-based chemotherapy. The serum tumor markers human beta-subunit chorionic gonadotropin (B-hCG) and alpha-fetoprotein (AFP) are extremely useful in both diagnosing disease and monitoring response to therapy.

Epidemiology

The average annual incidence rate for testicular cancer is 2.3 per 100,000 in the United States. Higher average annual rates are reported in Scandinavia, Switzerland, Germany, and New Zealand. There are approximately 5,500 new cases reported in the United States annually, with 350 expected deaths. This mortality rate is 10 times lower than 15 years ago, due largely to chemotherapeutic innovations to cure this disease. The cumulative lifetime risk of developing testicular cancer is 1 in 500. The incidence of testicular cancer in black Americans is one-fourth to one-third that of white Americans but 10 times higher than that of African blacks.

Age at Presentation

Testicular cancer occurs most frequently in young adult males aged 20-40. Other peaks occur in late adulthood (over 60 years) and in infancy (0-10 years).

Risk Factors

Genetics

A higher incidence of testicular tumors has been reported in twins, brothers, and family members, but there is not as yet a clear-cut genetic predisposition.

Cryptorchidism

The relative risk of testicular cancer in patients with cryptorchidism is roughly 5 to 10 times greater than that in the general population. Possible causative factors include abnormal germ cell morphology, elevated temperature, interference with blood supply, endocrine dysfunction and gonadal dysgenesis. Five to 10% of pa-

Urological Oncology, edited by Daniel Nachtsheim. ©2005 Landes Bioscience.

tients with cryptorchidism will develop a malignancy in the normally descended, contralateral testis.

Trauma

No clear relationship between testicular trauma and development of malignancy exists. However, trauma to an enlarged testis is usually the event that brings the patient to medical attention.

Hormones

Maternal exposure to diethylstilbestrol (DES) has been implicated in the pathogenesis of testicular maldescent and dysgenesis of the testis in male offspring. Whether the progeny of DES-treated mothers are predisposed to testicular tumors is still debatable.

Atrophy

Idiopathic and mumps-related atrophy of the testes have been suggested as possible causative factors.

Natural History

After malignant transformation, intratubular neoplasia (carcinoma in situ) grows beyond the basement membrane and may eventually replace some or all of the testicular parenchyma.

The tunica albuginea can prevent local invasion of the epididymis or spermatic cord; therefore, lymphatic or hematogenous spread may appear first. Tumors confined to the testis usually spread to retroperitoneal nodes, whereas involvement of the epididymis or cord may lead to pelvic and inguinal node metastases.

10

Classification and Pathology

Testicular tumors are of germ cell origin in 95% of cases. The non-germ cell tumors include Leydig cell tumors, gonadoblastoma, Sertoli-cell tumors, neoplasms of mesenchymal origin, adrenal rest tumors, carcinoid tumors and tumors metastatic to the testis (Table 10.1). Testicular germ cell tumors are divided into two major groups, seminomas (60%) and nonseminomas (40%).

The natural history and treatment of seminomas is drastically different from the nonseminomas and therefore accurate histologic diagnosis is essential. Seminomas are classified into three different groups based on histology: Classic (85%), anaplastic (10%) and spermatocytic (5%). The anaplastic seminomas are generally considered more aggressive and with more metastatic potential. The spermatocytic on the other hand have a lower metastatic potential, a more favorable prognosis and nearly half occur in men over the age of 50.

Histologically, classical seminomas are composed of bands or sheets of large cells with clear cytoplasm and densely staining nuclei surrounded by connective tissue stroma containing lymphocytes.

In contrast to seminomas, non-seminomas are histologically more heterogeneous. The World Health Organization has classified non-seminomas into five distinct histologic types: embryonal carcinoma, which accounts for approximately 50% of non-seminomas, mature and immature teratomas, choriocarcinomas and yolk sac tumors.

Table 10.1. Histologic classification of testis cancer

I. **Primary Neoplasms**
    A. Germinal neoplasms (demonstrating one or more of the following components:
        1. Seminoma
            a. Classic (typical seminoma)
            b. Anaplastic seminoma
            c. Spermatocytic seminoma
        2. Embryonal carcinoma
        3. Teratoma (with or without malignant transformation)
            a. Mature
            b. Immature
        4. Choriocarcinoma
        5. Yolk sac tumor (endodermal sinus tumor; embryonal adenocarcinoma of the prepubertal testis)
    B. Nongerminal neoplasms
        1. Specialized gonadal stromal neoplasms
            a. Leydig cell tumor
            b. Other gonadal stromal tumor
        2. Gonadoblastoma
        3. Miscellaneous neoplasms
            a. Adenocarcinoma of the rete testis
            b. Mesenchymal neoplasms
            c. Carcinoid
            d. Adrenal rest "tumor"
II. **Secondary Neoplasms**
    A. Reticuloendothelial neoplasms
    B. Metastases
III. **Paratesticular Neoplasms**
    A. Adenomatoid
    B. Cystadenoma of epididymis
    C. Mesenchymal neoplasms
    D. Mesothelioma
    E. Metastases

10

Seminomatous and non-seminomatous elements can be mixed and when this occurs, the tumor behaves as a non-seminoma.

Clinical Presentation

The most common presenting symptom is a painless testicular or scrotal mass. Thirty to 40% of patients may complain of a dull ache or heavy sensation in the lower abdomen, anal area or scrotum. Acute pain is the presenting symptom in 10% of patients. Up to 25% of patients may be initially misdiagnosed and treated for epididymitis. In 10% of patients, the presenting manifestations are those of metastatic disease including a neck mass or other lymphadenopathy, cough, back pain or nervous system disturbances. Gynecomastia can be seen in 5% of patients with testicular germ cell tumors due to systemic endocrine effects. Patient denial often contributes to several months delay in presentation.

Physical Examination

The scrotal exam should be performed with the patient standing and repeated with the patient supine. Bimanual examination of the normal contralateral testis is performed first. The relative size, contour and consistency of the normal testis compared to the suspicious testis are noted. The normal testis is homogeneous in consistency, freely movable and separable from the epididymis. Any firm, hard or fixed area within the substance of the tunica albuginea should be considered suspicious for malignancy until proven otherwise. The spermatic cord and scrotal skin should also be thoroughly examined. Physical examination should include examination of the abdomen to evaluate for palpable masses or visceral disease, a breast exam to evaluate for gynecomastia, and a thorough assessment of the presence of adenopathy, especially the supraclavicular (Virchow's) nodes.

Differential Diagnosis

Differential diagnosis of a testicular mass includes testicular torsion, epididymitis, epididymo-orchitis, hydrocele, spermatocele, hernia or hematoma.

Diagnosis

Transcrotal ultrasound is a rapid and reliable technique to confirm the presence of an intraparenchymal testicular mass, to rule out benign processes and effectively evaluate the contralateral testis. It should be employed in-patients with any suspicion of a testicular tumor (Figs. 10.1 and 10.2).

High inguinal (radical) orchiectomy with early clamping of the spermatic cord near the internal inguinal ring allows for complete removal of the primary tumor. Only in very rare circumstances are exploration and frozen biopsy performed and if done must be performed via an inguinal approach with temporary occlusion of the cord. Transcrotal orchiectomy is not performed in order to avoid contamination of the inguinal lymphatics. The pathology report should include the extent of the tumor (pathologic stage), whether there is vascular or lymphatic invasion, and the histologic composition of the lesion.

Tumor Markers

Testis germ cell tumors often produce marker proteins that are relatively specific and measurable in small quantities using radioimmunoassay techniques. These markers, especially B-hCG and AFP are clinically useful in the management of patients with germ cell tumors with respect to diagnosis, staging, monitoring response to therapy and predicting overall prognosis. Normally, the production of these oncofetal substances falls to nearly undetectable levels soon after birth. Their production by trophoblastic and syncytiotrophoblastic cells within germ cell neoplasms is the result of either the re-expression of repressed genes or the malignant transformation of a pluripotential cell that has retained the ability to differentiate into cells capable of producing oncofetal proteins.

Alpha-fetoprotein (AFP) is a 70,000 molecular weight (MW) single chain glycoprotein normally secreted by the fetal yolk sac. An elevated AFP may be associated with a number of malignancies (testis, liver, pancreas, stomach, and lung), normal pregnancy, benign liver disease, ataxia telangiectasia and tyrosinemia. AFP may be produced by pure embryonal carcinoma, teratocarcinoma, yolk sac tumor, or com-

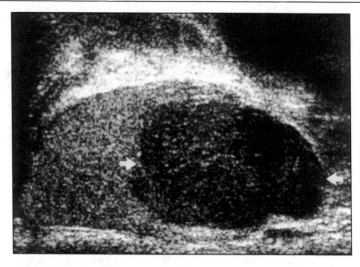

Fig. 10.1. Scrotal ultrasound of a 35-year-old man with the incidental finding of left testicular mass. Ultrasound shows a 2.3 x 2.0 x 1.7 cm hypoechoic well-defined mass within the left testicle. Tumor markers were negative. Pathology revealed pure seminoma.

10

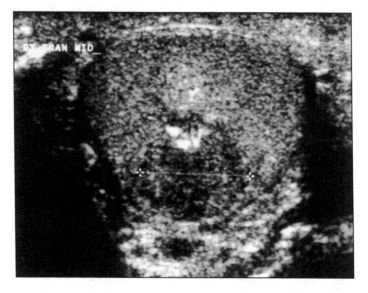

Fig. 10.2. Scrotal ultrasound of a 32-year-old man with the incidental finding of a right testicular mass. Ultrasound shows a 1.1 x 1.2 cm heterogeneous focal testicular mass with both hypoechoic and hyperechoic areas. Pathology revealed mixed germ cell tumor with embryonal and yolk sac elements.

bined tumor but not by pure choriocarcinorna or pure seminoma. The presence of an elevated AFP precludes the diagnosis of pure seminoma. The metabolic half-life of AFP in humans is between 5 and 7 days, which is useful in evaluating response to treatment.

Human chorionic gonadotropin (hCG) is a 38,000 MW double chain glyco-protein composed of alpha and beta subunits normally produced by the syncytiotrophoblastic cells of the placenta for maintenance of the corpus luteum. An elevated hCG may be seen in various other malignancies (liver, pancreas, stomach, lung, breast, kidney, and bladder) and in marijuana smokers. The alpha sub-unit closely resembles the alpha subunits of LH, FSH and TSH. Therefore, it is the antibodies to the beta subunit that are measured with the radioimmunoassay techniques. hCG is elevated in all patients with choriocarcinoma, 40-60% of embryonal carcinomas and 5-10 % of pure seminomas. The serum half-life of HCG is between 24 and 36 hours.

Lactate dehydrogenase (LDH) is a non-specific cellular enzyme with particularly high levels detectable in smooth, cardiac and skeletal muscles, liver, kidney and brain. It is not specific for testicular lesions but Isoenzymes I and II have been shown to correlate with tumor bulk, proliferation and death. It may have some role in monitoring patients with advanced seminoma and in marker negative nonseminoma patients with persistent disease. It must be correlated with other clinical findings in making therapeutic decisions.

Clinical Staging
Once germ cell neoplasm has been diagnosed by inguinal orchiectomy, clinical staging must be performed. Clinical staging is essential to aid the physician in constructing a treatment plan.

Staging Systems
Although many clinical and pathological staging systems exist, they all recognize three defined levels of tumor dissemination:

Stage I: Tumor confined to the testis or local structures

Stage II: Tumor metastatic to retroperitoneal lymph nodes

Stage III: Tumor metastatic beyond the retroperitoneum and/or to visceral organs (see Tables 10.2 and 10.3)

Sites of Metastases
The majority of testicular cancers spread through the lymphatics in an orderly fashion, although certain cell types such as choriocarcinoma metastasize through vascular dissemination. The primary lymphatic drainage from the right testis follows the course of the gonadal vessels primarily to the interaorto-caval lymph nodes and subsequently to precaval, pre-aortic and pericaval lymph nodes. Once the interaortal caval lymph nodes are involved, then tumor tends to spread to the left para-aortic area. The primary drainage of the left testis also follows the gonadal vessels to the left para-aortic nodes just below the level of the left renal vein and subsequently to pre-aortic nodes. Contralateral cross metastases do occur but more often in patients with right-sided tumors because of the lymphatic drainage from right to left. Inguinal metastases occur if the tunica albuginea has been invaded or

Table 10.2. Clinical staging systems for testis cancer

Boden/Gibb Stage	MSKCC	American Joint Committee
A (I)	A	TX unknown status
Tumor confined to testis		T0 no evidence primary
		T1 confined to testis
		T2 beyond tunica
		T3 invades rete testis of epididymis
		T4a invades cord
		T4b invades scrotum
B (II)	B1 <5 cm	
Spread to regional nodes	B2 >5 cm	
	B3 >10 cm ("bulky")	
C (III)	CIII	
Spread beyond		
retroperitoneal nodes		

when previous surgery such as inguinal herniorrhaphy or orchiopexy has altered the normal lymphatic flow. Distant metastatic spread is usually to the lungs with intraparenchymal pulmonary involvement. Subsequent spread may be to other viscera including liver, brain, or bone.

Imaging Studies

Included in clinical staging are posteroanterior and lateral chest x-rays to assess lung parenchyma and mediastinal structures and an abdominal computerized tomography (CT) scan to identify retroperitoneal lymph node involvement (Fig. 10.3).

Lymphangiography was used in the past to determine the extent of retroperitoneal involvement but this time consuming technique has largely been replaced by CT scanning.

Treatment

Seminoma

Seventy-five to 80% of seminomas are clinical stage I at the time of diagnosis, 10-20% are stage II, and less than 5% are clinical stage III. Prognosis worsens with clinical stage. The staging of seminoma following inguinal orchiectomy should include a thorough physical examination, chest x-ray, CT of chest, abdomen and pelvis, as well as post-orchiectomy determination of hCG. If there was pre-orchiectomy elevation of hCG, weekly serum hCG levels should be obtained until the hCG normalizes. If the serum hCG remains elevated in the face of a negative metastatic work-up, it must be assumed that the patient has metastatic stage II or III disease.

Stage I

In patients who have seminoma, a negative metastatic workup and normalization of serum hCG following orchiectomy is considered stage I. Adjuvant radiation therapy with a dose of 2,500-3,000 rads (25-30 Gy) is then administered to the retroperitoneal lymph node groups including the ipsilateral external iliac, the bilateral common iliac, the paracaval and the paraaortic nodes. The orchiectomy scar is also irradiated.

Table 10.3. TNM staging of testis tumors

Primary Tumor

pTx	Primary tumor cannot be assessed (if no orchiectomy has been performed,
Tx	is used)
pPT0	No evidence of primary tumor (e.g., histologic scar in testis)
pTis	Intratubular germ cell neoplasms (carcinoma in situ)
pT1	Tumor limited to the testis and epididymis and no vascular-lymphatic invasion: tumor may invade into the tunica albuginea but not the tunica vaginalis
pT2	Tumor limited to the testis and epididymis with vascular-lymphatic invasion of tumor extending through the tunica albuginea with involvement of tunica vaginalis
pT3	Tumor invades the spermatic cord with or without vascular-lymphatic invasion
pT4	Tumor invades the scrotum with or without vascular-lymphatic invasion

Regional Lymph Nodes

Clinical Involvement

Nx	Regional lymph nodes cannot be assessed
N0	No regional lymph node metastasis
N1	Lymph node mass 2 cm or less in greatest dimension; or multiple lymph nodes, none more than 2 cm in greatest dimension
N2	Lymph node mass more than 2 cm but not more than 5 cm in greatest dimension; or multiple lymph nodes, any one mass more than 2 cm but not more than 5 cm in greatest dimension
N3	Lymph node mass more than 5 cm in greatest dimension

Pathologic Involvement

pN0	No evidence of tumor in lymph nodes
pN1	Lymph node mass 2 cm or less in greatest dimension and 5 nodes or less positive, none more than 2 cm in greatest dimension
pN2	Lymph node mass more than 2 cm but not more than 5 cm in greatest dimension; more than 5 nodes positive, none more than 5 cm; evidence of extranodal extension of tumor
pN3	Lymph node mass more than 5 cm in greatest dimension

Distant Metastases

M0	No evidence of distant metastases
M1	Nonregional modal or pulmonary metastases
M2	Nonpulmonary visceral metastases

Serum Markers

	LDH	hCG (mIU/mL)	AFP (ng/mL)
S0	Marker study levels within normal limits		
S1	<1.5 XN	<5,000	<1,000
S2	1.5-10 XN	5,000-50,000	1,000-10,000
S3	>10XN	>50,000	>10,000

---

continued on next page

Patients with clinical stage I seminoma, treated with radical orchiectomy and radiation therapy achieve nearly a 100% five-year disease free survival.

Stage IIA

For those patients with low volume stage II disease orchiectomy followed by radiation therapy results in approximately 70-90% five-year survival.

Table 10.3. Continued

Stage Grouping
Stage I
1A T_1, N_0, M_0, S_0
1B T_2, N_0, M_0, S_0
        T_3, N_0, M_0, SO
        T_4, N_0, M_0, S_0
IS   Any T, N_1, M_0, and S
Stage II
IIA   Any T, N_1, M_0, S_0
        Any T, N_1, M_0, S_0
IIB   Any T, N_2, M_0, S_0
        Any T, N_2, M_0, S_1
IIC   Any T, N_3, M_0, S_0
        Any T, N_3, M_0, S_1
Stage III
IIIA   Any T, any N, M_1, S_0
        Any T, any N, M_1, S_1
IIIB   Any T, any N, M_0, S_2
        Any T, any N, M_1, S_2
IIIC   Any T, any N, M_0, S_3
        Any T, any N, M_1, S_3
        Any T, any N, M_2, any S

N indicates the upper limit of normal for the LDH assay. Adapted with permission from the American Joint Committee on Cancer (AJCC®), Chicago, IL. From AJCC® Cancer Staging Manual, 5th ed. Philadelphia: Lippincott-Raven Publishers, 1997.

10

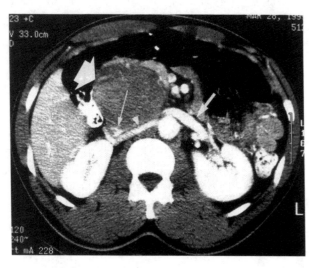

Fig. 10.3. CT of the abdomen of a 48-year-old man with right testis seminoma. Tumor markers were negative. CT shows extensive retroperitoneal lymphadenopathy (wide arrow) compressing the IVC (thin arrow) (Arrowhead: right renal artery; medium arrow: left renal vein).

Stage IIB or Higher (Stage III)

For those patients with bulky stage II disease (IIB) or stage III, orchiectomy and radiation afford cure in only 25-40% of patients treated, and approximately one-third of patients will develop metastatic disease outside of the treated fields. For these reasons, platinum-based chemotherapy is recommended followed by RPLND (or radiation therapy) if prior treatment has failed. The combination of cis-platin, etoposide or vinblastine and bleomycin (BEP) chemotherapy has been found to be quite effective. Eighty to 90% of patients treated with platinum-based chemotherapy will achieve complete remission and five-year disease-free survival.

Non-Seminoma

The prognosis for nonseminoma is less favorable than for seminoma. Approximately 50-70% of patients with non-seminoma present with metastatic disease at the time of diagnosis.

Stage I

For clinical stage I nonseminoma with normalization of serum markers following orchiectomy, a retroperitoneal lymph node dissection (RPLND) is generally recommended. RPLND can be performed via the transabdominal, thoracoabdominal or laparoscopic approach. If serum markers, CT scan and laparotomy are all negative, a modified, bilateral RPLND can be performed. Expectant observation with monthly follow-up may be possible in a select group of very reliable patients. However, the cost-effectiveness of this protocol must be taken into account.

If the patient has pathologic stage I non-seminoma following retroperitoneal lymphadenectomy, no further treatment is warranted. The cure rate for pathological stage I disease is roughly 95% with surgery alone. The 5-10% of patients who relapse usually do so within the first two years and have a high cure rate approaching 100% with salvage chemotherapy. Relapse following negative RPLND is usually to the lungs and rarely to the retroperitoneum.

Twenty to 25% of patients with clinical stage I disease will be found to be understaged at time of RPLND.

RPLND is generally well-tolerated with low associated morbidity (5-25%, mostly from atelectasis, ileus, and lymphocele) and minimal reported mortality rates (1%). The major long-term complication following RPLND is ejaculatory dysfunction. Approximately 50-60% of men are found to be subfertile at the time of diagnosis of nonseminomatous germ cell tumor. As therapy for testis cancer may have potential detrimental effects on spermatogenesis, early counseling and sperm cryopreservation should be encouraged.

Prior to the development of the modified RPLND, the lymphadenectomy included a complete bilateral dissection from above the renal hilar area, extending down to where the ureters cross the common iliac arteries. Stimulation of the T12-L3 sympathetic nerve fibers during ejaculation results in seminal vesicle, prostatic and vasal contraction, as well as bladder neck closure. The disruption of these sympathetic nerve fibers resulted in a high incidence of infertility either due to failure of seminal emission or from retrograde ejaculation.

The modification of the surgical boundaries for the RPLND has allowed for preservation of ejaculation in greater than 90% of patients.

10

Stage II and III

For low stage II nonseminomatous germ cell tumors as with stage I tumors, surgical excision is the mainstay of treatment. Controversy exists for stage II disease as to whether surgery is sufficient treatment or if adjuvant BEP chemotherapy should be given. The general consensus is that for IIA disease, observation after RPLND is sufficient. With IIB tumors, RPLND should be followed by two courses of BEP. For very advanced stage II disease as well as stage III disease, four cycles of BEP followed by post-chemotherapy RPLND is recommended with cure rates of greater than 60%.

High dose chemotherapy and autologous bone marrow transplant are being investigated for use in patients with tumors having poor prognostic factors.

Surveillance and Follow-Up

Routine follow-up for these patients includes chest x-ray, abdominal and pelvic CT scan, and serum tumor markers. These patients should be re-evaluated every 3 months for the first year and every six months for an additional 3-4 years with these studies.

Other Testis Neoplasms

These tumors comprise 5-10% of all testis tumors. They include Leydig cell tumors, Sertoli cell tumors and gonadoblastoma and are generally treated and cured with inguinal orchiectomy.

Selected Readings

1. Mostofi FK. Testicular tumors: Epidemiologic, etiologic and pathologic features. Cancer 1973; 32:1186-1201.
2. Cosgrove MD, Benton B, Henderson B. Male genitourinary abnormalities and maternal diethylstilbestrol. J Urol 1977; 117:220-222.
3. Richie JP. Neoplasms of the testis. In: Walsh PC, Retik AL, Vaughan ED, Wein AJ, eds. Campbell's Urology. 7th ed. Philadelphia: W.B. Saunders, 1997.
4. Vogelzang NJ, Lange PH. Tumors of the testes. In: Moosa et al, eds. Comprehensive Textbook of Oncology. 2nd ed. Baltimore: Williams and Wilkins, 1991.
5. Malkowicz SB, Wein AJ. Adult genitourinary cancer. In: Hanno PM, Wein AJ, eds. Clinical Manual of Urology. 2nd ed. Philadelphia: McGraw-Hill, Inc., 1993.

10

CHAPTER 11

Pediatric Genitourinary Cancer

Christopher S. Cooper and Howard M. Snyder III

Introduction

Active research and advances in treatment of pediatric genitourinary cancers result in a dynamic field for study. The pediatric cancers covered in this chapter include Wilms' tumor as well as other kidney tumors, neuroblastoma, rhabdomyosarcoma, and prepubertal testicular tumors. For each cancer the incidence, etiology, pathology, presentation, evaluation, treatment, and outcomes will be discussed.

Wilms' Tumor

Incidence

Wilms' tumor is the most common malignant neoplasm of the urinary tract in children. This tumor comprises 8% of all solid tumors in children. The peak incidence occurs between 3 and 4 years of age and 90% occur in children younger than 7 years. Boys and girls are almost equally affected by Wilms' tumor.

Etiology

Wilms' tumor originates from abnormal renal histogenesis. Nephrogenic rests or nephroblastomatosis may be precursor lesions of a Wilms' tumor. A nephrogenic rest consists of a focus of abnormally persistent nephrogenic cells, and a second factor may transform the rest into a malignancy. Nephroblastomatosis consists of multifocal or diffuse nephrogenic rests. Most rests undergo regression and become sclerotic or obsolescent. Some become hyperplastic but may regress later. Unfortunately, a dormant, hyperplastic, or regressing rest maintains neoplastic potential.

The microscopic appearance of Wilms' tumor and nephrogenic rests can be indistinguishable. The diagnosis of Wilms' tumor may depend on the shape and growth characteristics of the lesion as determined by serial imaging studies. With hyperplasia a rest tends to maintain its original shape, unlike a neoplasm which is often spherical.

The Wilms' tumor 1 gene (WT1) is located in the 11p13 region and is critical for normal genitourinary development. Mutations are associated with the development of Wilms' tumor as occurs in the Denys-Drash or WAGR syndromes. Other gene mutations distal to the WT1 gene in the 11p15 (WT2) region are associated with Wilms' tumors occurring in the Beckwith-Wiedemann syndrome.

Pathology

Wilms' tumor frequently consists of an encapsulated solitary tumor within the kidney. Necrosis within the tumor is common. Multiple lesions may occur with

---

Urological Oncology, edited by Daniel Nachtsheim. ©2005 Landes Bioscience.

nephroblastomatosis. Invasion of the renal vein by the tumor occurs up to 20% of the time. The lymph nodes are frequently enlarged without metastatic disease, making gross assessment of nodal involvement unreliable.

Microscopically, the tumor demonstrates a triphasic histology consisting of blastemal, epithelial, and stromal cells. The stromal component may differentiate into striated muscle, cartilage, or rarely, fat or bone. The epithelial component varies from well-differentiated resembling mature tubules to a very primitive appearance.

The histologic demonstration of anaplasia portends a worse prognosis. Anaplasia consists of a three-fold variation in nuclear size with hyperchromatism and abnormal mitotic figures. Anaplasia occurs more frequently in older children. Diffuse distribution of anaplasia throughout the lesion or at any extrarenal site conveys a worse prognosis than focal anaplasia.

Presentation

Most children with Wilms' tumor present with an abdominal mass (Fig. 11.1). The differential diagnoses of benign childhood renal masses are presented in Table 11.1. About 30% of patients present with abdominal pain, but most appear well and are asymptomatic. Tumor hemorrhage or rupture may present with an acute abdomen.

Physical examination reveals a firm, non-tender, smooth mass which rarely crosses the midline. Hypertension occurs in up to 63% of children with Wilms' tumor. With tumor propagation into the vena cava, a child may develop a varicocele or even congestive heart failure.

Associated Anomalies

Congenital abnormalities occur in about 15% of children with Wilms' tumor, as shown in Table 11.2.

11

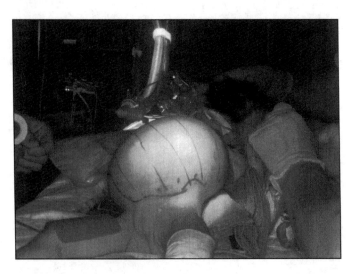

Fig. 11.1. Preoperative image of child with Wilms' tumor creating a large abdominal mass.

Table 11.1. Differential diagnoses of benign childhood renal masses

- Hydronephrosis
- Congenital mesoblastic nephroma
- Multicystic-dysplastic kidney
- Abscess
- Xanthogranulomatous pyelonephritis
- Multilocular cyst
- Polycystic kidney
- Glomerulocystic kidney
- Angiomyolipoma
- Teratoma
- Pseudotumor

Table 11.2. Wilms' tumor-associated anomalies

- Sporadic (non-familial) aniridia: one-third develop Wilms' tumor
- WAGR syndrome (Wilms' tumor, aniridia, GU abnormalities, mental retardation)
- Hemihypertrophy: 3% develop Wilms' tumor
- Beckwith-Wiedemann syndrome (gigantism, macroglossia, omphalocele, genitourinary anomalies): 10% develop tumor
- Denys-Drash (ambiguous genitalia, progressive renal failure, Wilms' tumors)
- Musculoskeletal
- Dermatologic (hemangiomas, multiple nevi, café-au-lait spots)
- Genitourinary (renal anomalies, hypospadias, cryptorchidism)

Evaluation

Wilms' tumor characteristically demonstrates a heterogeneous echo pattern on ultrasonography and can vary from predominately cystic to solid. MRI accurately evaluates the extent and size of the Wilms' tumor, which frequently gives variable signal intensities. CT also provides precise anatomic delineation of the renal and retroperitoneal anatomy (Fig. 11.2). Despite the accuracy with these techniques, false negative results do occur and a contralateral renal exploration by opening Gerota's fascia and inspecting all surfaces of the kidney is still required since bilateral Wilms' occurs in close to 10% of children.

Treatment

Effective treatment of Wilms' tumor relies on the appropriate use of surgery, radiation, and multi-agent chemotherapy, and varies with stage and histology. The staging system employed by the National Wilms' Tumor Study (NWTS) group is outlined in Table 11.3.

Surgery

Prior to removal of the primary tumor, the extent of the tumor is evaluated, as well as the liver and lymph nodes. The contralateral kidney must be explored by opening Gerota's capsule and inspecting the entire surface at the time of surgery. Biopsy of any suspicious lesion is required.

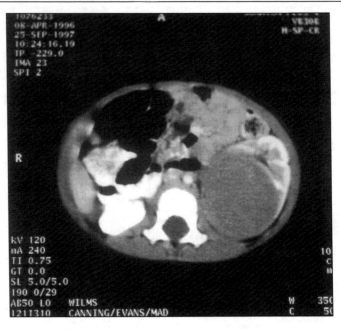

Fig. 11.2. CT scan demonstrating a large left-sided Wilms' tumor.

If the primary tumor invades surrounding organs and prevents safe removal, it is prudent to wait until after treatment with chemotherapy and/or radiation. When the kidney can be removed it is taken along with the adrenal gland if the tumor involves the upper pole (Fig. 11.3). Tumor rupture with diffuse spillage increases the chance of abdominal relapse. Preoperative radiation in addition to chemotherapy will decrease this risk.

Chemotherapy

Preoperative chemotherapy can be used selectively and is appropriate for extensive tumors, tumors with major vena cava invasion, or with bilateral tumors. Actinomycin D, vincristine, and adriamycin all have been effective in the treatment of Wilms' tumor. The use of these agents, as well as radiotherapy varies with the tumor stage and histology.

A summation of the current treatment for Wilms' tumor stages I-IV employed by NWTS-V is outlined in Table 11.4. The treatment of bilateral disease (stage V) employs a biopsy followed by preoperative chemotherapy and bilateral partial nephrectomies. When metastatic or recurrent disease is encountered, doxorubicin is routinely given if the patient has only received dactinomycin and vincristine. For those who have already received three-agent chemotherapy, there is no well-established regimen.

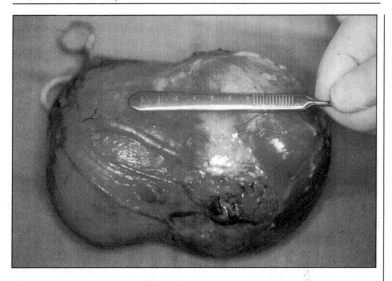

Fig. 11.3. Nephrectomy specimen of kidney with large Wilms' tumor.

Table 11.3. Wilms' tumor stages

Stage I: Tumor within the kidney and completely excised.
Stage II: Tumor beyond the kidney and completely excised. May have invaded vessels or a local confined spill may have occurred.
Stage III: Tumor left within the abdomen and can include positive lymph nodes, positive surgical margins, peritoneal metastases, diffuse unconfined tumor spillage.
Stage IV: Metastatic disease via hematogenous route.
Stage V: Bilateral Wilms' tumor.

Table 11.4. Summary of treatment of Wilms' stage I-IV

Stage I: Dactinomycin and vincristine
Stage II: Favorable histology: as with stage I
Unfavorable histology: doxorubicin, cyclophosphamide, etoposide, vincristine, and radiation
Stage III and IV:
Favorable histology: dactinomycin, vincristine, doxorubicin, and radiation
Unfavorable histology: as with stage II unfavorable

11

Radiation

The routine use of radiation in the treatment of Wilms' tumor is listed in Table 11.4. Radiation to the flank of children occurs 1 to 3 days following surgery. Supplemental radiation is directed to regions with residual tumor. Pulmonary metastases result in radiation to both lungs regardless of the number or location of metastases.

Bilateral Wilms' Tumor

Synchronous bilateral Wilms' tumors occur in about 6% of children with Wilms' tumor and metachronous tumors occur in another 1%. Nephrogenic rests occur in all patients with bilateral Wilms' tumors and frequently present at an earlier age. Anaplasia is less frequent in children under the age of 2 with bilateral Wilms' tumors and, accordingly, the younger children tend to have a better prognosis. They have more genitourinary anomalies and a higher incidence of hemihypertrophy than children with unilateral Wilms' tumor.

The overall survival of patients with synchronous bilateral Wilms' tumors is 76% at 3 years. This is in contrast to those who develop a metachronous bilateral Wilms' tumor, in which only 39% are free of disease 2 years after developing the second Wilms' tumor. The histology, stage of the most advanced lesion, and presence of lymph node metastases are important prognostic indicators with bilateral Wilms' tumors.

A correct preoperative diagnosis of bilateral Wilms' tumor is made about two-thirds of the time and a contralateral kidney exploration makes the diagnosis in the other third. The treatment following diagnosis by radiographic imaging and/or biopsy, is chemotherapy and then surgery. This approach frequently permits a renal-sparing approach with partial nephrectomy or excisional biopsy and at times the larger tumor responds better to chemotherapy than the smaller one. Nephron-sparing surgery is contraindicated when diffuse anaplasia is encountered. Bilateral nephrectomy and renal transplantation may be appropriate for those with bilateral, diffuse anaplasia.

Outcomes

The 4-year post-nephrectomy survival results from the National Wilms' Tumor Study III are listed in Table 11.5. The most important prognostic factor is histology. Diffuse anaplasia is an unfavorable histology. Approximately 5% of children with Wilms' tumor have anaplasia, and this occurs more frequently in the older child. When anaplasia is well-circumscribed and focally contained within the primary tumor, it is not considered unfavorable.

Hematogenous metastases also worsen the child's prognosis and are present at diagnosis in 11-15% of patients. Metastatic disease to the lungs occurs most often (85%), followed by metastases to the liver, bone, or brain. Metastases to the lymph nodes also predict a worse outcome with increased risk of local recurrence, and thus, abdominal radiotherapy is added to chemotherapy. A tumor with renal vasculature invasion carries a higher risk of local relapse.

Other Pediatric Kidney Tumors

The most common solid renal tumor of the neonate is the congenital mesoblastic nephroma (CMN). Polyhydramnios occurs in association with this tumor, which

11

Table 11.5. Wilms' tumor four-year survival (%) after nephrectomy and chemotherapy ± radiotherapy

	Histology	
Stage	Favorable	Unfavorable
I	97	68
II	92	55
III	87	45
IV	83	4

most often affects boys. It is a firm tumor grossly resembling a leiomyoma. Histologic evaluation demonstrates sheets of spindle-shaped uniform cells with a fibroblastic appearance. Treatment involves complete excision of this benign lesion with a nephrectomy.

The multilocular cyst is a round and smooth tumor with multiple cysts. The septa of the cysts in the pediatric tumor are composed of fibrous tissue and may contain well-differentiated tubular structures. It is usually unilateral and the peak incidence occurs in the pediatric age group, usually affecting boys. Treatment requires local excision; however, the kidney may be preserved. The second peak incidence of multilocular cyst occurs in the adult and usually in women. The cystic, partially differentiated nephroblastoma mimics the cystic nephroma but the septae contain blastema. Frozen-section analysis is essential to plan surgical treatment since a nephrectomy is indicated for treatment of the cystic, partially differentiated nephroblastoma.

Table 11.6 describes malignant pediatric renal neoplasms other than Wilms' tumor. The rhabdomyosarcoma tumor of the kidney is a Wilms' tumor variant. This tumor, unlike the rhabdoid tumor, has a favorable outcome. The rhabdomyosarcoma contains fetal striated muscle that distinguishes it from the rhabdoid tumor, which does not contain muscle.

Neuroblastoma

Incidence
Neuroblastoma is the most common malignant tumor of infancy and second most common malignant solid tumor of childhood. Boys are slightly more affected than girls (1.1:1). One-half of the cases of neuroblastoma occur in children under the age of 2, and by age 4 more than 75% of the cases of neuroblastoma have occurred. It arises from cells of the neural crest.

Etiology and Pathology
The clinical behavior is variable and likely reflects a number of different genetic abnormalities that determine the tumor phenotype. A hypothesized neuroblastoma suppressor gene is located on the short arm of chromosome 1. A ganglioneuroma is thought to arise from a similar cell line as the neuroblastoma but is a benign lesion. On occasion, a neuroblastoma undergoes spontaneous maturation to a benign ganglioneuroma. A ganglioneuroblastoma can behave in either a benign or malignant manner.

11

Table 11.6. Primary pediatric renal cancers

Renal Cell Carcinoma: Most common primary renal neoplasm of childhood other than Wilms' tumor
- Presentation: hematuria, flank pain, mass, weight loss
- Metastasis: lymph nodes, bones, lungs, liver
- Increased with von Hippel-Lindau and PCKD
- Survival: 50%
- Treatment: radical nephrectomy

Rhabdoid Tumor: Large cells with eosinophilic inclusions like rhabdomyoblasts but no muscle
- Metastatic: brain
- Very lethal (survival: 18%)
- Treatment: no clearly effective chemotherapy yet defined

Clear Cell Sarcoma
- Metastasis: bone and brain
- Predominantly male infants
- Treatment: chemotherapy with adriamycin improves prognosis

Rare Cancers
- Rhabdomyosarcoma (unlike rhabdoid tumor it contains fetal striated muscle)
- Neuroblastoma
- Leiomyosarcoma
- Transitional cell carcinoma

The tumor frequently forms a pseudocapsule and tends to infiltrate through this capsule. Multiple small, round cells resembling lymphocytes or neuroblasts make up the neuroblastoma which is considered one of the "small, blue, round cell" tumors of childhood. With well-differentiated tumors, the cells form into rosettes and neurofibrils. These rosettes can be seen on bone marrow aspirate of marrow infiltrated with neuroblastoma.

Presentation

Neuroblastoma occurs in multiple sites along the sympathetic chain from head to pelvis. Two-thirds of abdominal neuroblastomas originate in the adrenal gland and these metastasize less frequently than nonadrenal abdominal neuroblastomas. A variety of presentations result from the multiple possible tumor locations. A mass is the most common presenting symptom. Neuroblastoma is characteristically fixed, irregular, firm, non-tender, and extends beyond the midline. Neuroblastomas arising from an abdominal paravertebral sympathetic ganglion have the potential to grow through an intervertebral foramen and compress the spinal cord. Occasionally, tumor secretion of catecholamines creates paroxysmal attacks of sweating, pallor, headaches, hypertension, palpitations, and flushing. Table 11.7 compares Wilms' tumor with neuroblastoma.

Seventy percent of patients have metastases by the time of diagnosis. Metastases in the younger child tend to occur in the liver, and in the older child arise more often in the bone. Manifestations of disseminated disease include fever, malaise, anorexia, weight loss, irritability, bone pain, and pallor secondary to anemia. These symptoms create an unwell-appearing child. Over half of children with neuroblastoma have

Table 11.7. Comparison of Wilms' tumor and neuroblastoma

Wilms' Tumor	Neuroblastoma
Frequent associated congenital anomalies	Low incidence of associated anomalies (except 2% incidence of brain and skull defects)
Presents in well child	Presents in sick child
Usually unilateral mass	Often mass crosses midline
Peak age 3-4 years	Peak age 22 months
Arises in the kidney	Primary renal neuroblastoma very rare
Occasional calcification	"Stippled" calcification on plain films in 50%, and 80% by CT

bone marrow metastases and may not demonstrate radiographic changes in the bone. Metastases to the periorbital region may cause periorbital ecchymoses, edema, and proptosis. Disseminated metastatic disease occurs much more frequently in children over 1 year of age.

Subcutaneous metastatic nodules frequently present in infants. The nodules often have a bluish "blueberry muffin" appearance. When these nodules occur in combination with hepatomegaly and tumor cells in the bone marrow without bone lesions, the child is classified as having stage IV-S disease. Children with stage IV-S disease have a strong likelihood of spontaneous regression and are considered to have "favorable disease" with a survival rate of 80%.

Evaluation

The history and physical examination provide clues as to whether the child may have a localized favorable tumor or a metastatic unfavorable tumor. A healthy-appearing infant with an asymptomatic mass is likely to have favorable disease. An older child who presents with constitutional symptoms likely suffers from disseminated metastases. A bone marrow aspiration from each of the posterior iliac crests constitutes a required part of the evaluation of neuroblastoma. The aspirate can be done at the time of primary surgery for a child who likely has favorable disease. In the child with unfavorable disease, a preoperative bone marrow aspiration may prevent the need for an open tissue biopsy.

The two major metabolites of catecholamine production by neuroblastoma are vanillylmandelic acid (VMA) and homovanillic acid (HVA). These metabolites are elevated in the urine in 95% of patients with neuroblastoma. Imaging studies, which have proved useful and are recommended in the evaluation of neuroblastoma, include a CT scan and/or MRI, chest radiography, skeletal survey, and 1-meta-iodobenzylguanidine (MIBG) radionuclide scanning.

Staging
The staging of neuroblastoma is outlined in Table 11.8.

Treatment and Outcome

The appropriate use of surgery, chemotherapy, and radiotherapy depends on the patient's prognostic factors. Prognoses may be grouped into "favorable" or "unfavorable" according to age, stage, tumor markers, and biological factors, including histology. Stage is the most important prognostic factor and age at diagnosis is the only

11

Table 11.8. International Neuroblastoma Staging System (INSS)

Stage 1: Tumor within area of origin completely excised, with or without microscopic residual disease.
Stage 2A: Tumor does not cross midline and incomplete gross resection.
Stage 2B: Tumor does not cross midline, with ipsilateral positive lymph nodes.
Stage 3: Tumor crosses midline, and may involve regional lymph nodes bilaterally.
Stage 4: Distant metastases except as defined in stage 4S.
Stage 4S: Localized stage 1 or 2 and spread to liver, skin, or bone marrow (<10% of nucleated marrow cells are malignant and no bony lesions on skeletal survey).

other independent clinical prognostic factor. Stages 3 or 4 and age over a year of life are both considered as unfavorable prognostic factors. Elevated ferritin, lactic dehydrogenase or neuron-specific enolase levels portend a worse prognosis. Histologic characteristics considered unfavorable include poorly differentiated neuroblasts, a nodular stroma, and increased mitotic activity. The MYCN oncogene is frequently amplified with tumors containing unfavorable histology. A diploid DNA content is also confers a worse prognosis.

The initial surgery serves to establish the diagnosis and stage of the tumor, provide tumor tissue for evaluation, and excise the tumor if possible without undue morbidity. A staging lymph node sampling is performed. A second surgery may remove residual disease. Chemotherapeutic agents used for neuroblastoma include cyclophosphamide, doxorubicin, etoposide, and platinum-based agents. Radiation may be used prior to bone marrow transplant as an ablative method, as well as for control of local and metastatic disease.

Most patients (95%) with favorable disease (stages 1 and 2) can be cured with surgical excision alone. Subsequent surgery is performed for local recurrence. Patients with stage 4S disease have a high rate of spontaneous regression and require no treatment if they are asymptomatic.

Children at intermediate risk include those infants with International Neuroblastoma Staging System (INSS) stages 3 and 4 that lack MYCN amplification, as well as children with INSS stage 3 tumors containing favorable biological features. These children require multi-agent chemotherapy and surgery, and survival ranges from 55-90%.

Children at high risk include those over 1 year of age with stage 3 tumors and unfavorable biology as well as all children with stage 4 disease. Also included in the high-risk group are infants with stages 3, 4, or 5-S and MYCN amplification as well as older children with stage 2 and unfavorable histology. Survival in this group ranges from <10% to 30% despite aggressive multi-modal therapy.

Therapeutic approaches aimed at improving the outcome of the high-risk patients include bone marrow transplant following a combination of chemotherapy and total body irradiation. Targeted radiation with labeled MIBG or monoclonal antibodies to treat neuroblastoma is also being evaluated. Gene therapy remains as a possible future therapeutic option.

Rhabdomyosarcoma

Incidence
Sarcomas are the fifth most common solid tumor in children after CNS tumors, lymphomas, neuroblastomas, and Wilms' tumor. Rhabdomyosarcoma accounts for half of all soft tissue sarcomas in children. It affects boys more often than girls (3:1). Most cases occur between 2 and 6 years of age.

Etiology and Pathology
Rhabdomyosarcomas arise from any part of the body that contains embryonal mesenchyme. This normally differentiates into skeletal muscle. As many as 20% of these tumors are from the genitourinary tract and 10% arise from pelvic or retroperitoneal sites. The most common genitourinary sites involve the prostate, bladder (usually in the area of trigone), and anterior vaginal wall. Other sites of rhabdomyosarcoma that present to the urologist include paratesticular and pelvic locations.

Rhabdomyosarcoma grows rapidly and invades adjacent tissue, making complete surgical excision difficult despite a deceptively well-circumscribed appearance. It spreads by both the lymphatic and vascular systems. When patients develop metastases, 80% of them are evident within 1 year of diagnosis.

Several histologic types of rhabdomyosarcoma have been described and include embryonal, alveolar, and pleomorphic. Embryonal rhabdomyosarcoma accounts for 60% of cases and occurs in younger children. Sarcoma botryoides is a polypoid type of embryonal rhabdomyosarcoma that tends to arise in hollow organs such as the bladder or vagina and resembles a cluster of grapes. Histologically, embryonal rhabdomyosarcoma resembles developing skeletal muscle from a 7- to 10-week old fetus.

Presentation
As with neuroblastoma, the multiple possible sites of rhabdomyosarcoma create a variety of presentations. A large palpable mass is the single most common presentation. With involvement of the bladder or prostate, the child may present with lower urinary tract obstruction or hematuria. With sarcoma botryoides the tumor may prolapse through the urethra or vagina. A pelvic tumor may grow large before it produces symptoms by impinging on the rectum or genitourinary tract.

11

Evaluation
Routine evaluation should employ chest x-rays, CT or MRI of the primary tumor, bone marrow examination, as well as routine blood tests. Cystoscopy and vaginoscopy may help delineate the site of tumor. A histologic diagnosis is established by biopsy. Since 20% of genitourinary rhabdomyosarcomas have spread to the retroperitoneal lymph nodes by the time of diagnosis, an inspection and sampling of these nodes is required to accurately stage the tumor.

Staging
The Intergroup Rhabdomyosarcoma Study (IRS) Committee currently employs a modification of the TNM staging system, demonstrated in Table 11.9.

Table 11.9. TNM rhabdomyosarcoma staging

T1: Tumor confined to organ tissue of origin
  a. Tumor <5 cm in greatest diameter
  b. Tumor >5 cm
T2: Tumor outside organ tissue of origin
  a. Tumor <5 cm in greatest diameter
  b. Tumor >5 cm
N0: Regional nodes not clinically involved
N1: Regional nodes clinically involved
M0: No distant metastases
M1: Metastases present

Treatment

Historical treatment with surgical excision alone resulted in a 40% survival rate for patients with vaginal tumors and 73% for patients with bladder or prostate rhabdomyosarcoma. Survival rates have improved with the use of multi-modal therapy. The survival results with genitourinary tumors are better than for rhabdomyosarcoma in other parts of the body, and may be related to earlier symptoms and detection. Radiation and chemotherapy are effective treatment modalities. Because late recurrences can occur following apparent cures with chemotherapy alone, either adjunctive radiation or surgery is directed toward the primary site.

Current treatment usually begins after tumor biopsy and consists of 8 weeks of treatment with vincristine, actinomycin D, and cyclophosphamide (VAC). Even patients who appear to have had the tumor completely excised receive vincristine and actinomycin D. If no response occurs after treatment with VAC, conservative surgical excision and marking of the tumor margins is performed followed by the addition of radiotherapy and chemotherapy. Surgical excision of gross residual disease following chemotherapy may demonstrate no viable tumor cells and spare the patient from radiotherapy. An alternative approach for patients with residual disease after treatment with primary chemotherapy is radiotherapy followed by surgical excision of any remaining tumor; however, the pathologic analysis of tissue removed following radiotherapy is difficult to interpret.

Outcome

Using a combination of primary chemotherapy and radiation improves both survival and bladder salvage. The five-year survival rates in patients with local disease in IRS group I equaled 93%. The survival rates for those in group II and III equaled 81% and 74%, respectively. At present, about one-half of the patients with bladder-prostate tumors retain their bladders.

Conservative surgery after chemotherapy and radiation also has produced excellent disease-free survival for patients with vaginal and vulvar tumors. The outlook for patients with metastatic disease remains dismal (long-term survival rates are only 20%).

11

Table 11.10. Prepubertal testis tumor types and distribution

- Yolk sac: 62%
- Teratoma: 15%
- Gonadal stroma (Leydig/Sertoli/juvenile granulosa cell, undifferentiated): 9%
- Epidermoid cyst: 2%
- Gonadoblastoma: 1%
- Others (benign tumors of supporting tissue, sarcomas, lymphoma, leukemia): 11%

Prepubertal Testicular Tumors

Incidence

Testicular tumors account for 2% of all pediatric solid tumors. The peak incidence of pediatric testicular tumors is in 2-year-olds. Germ cell tumors account for 60-77% of testicular tumors in children, whereas 95% of adult testicular tumors are germ cell tumors.

Classification

Table 11.10 illustrates various types and relative frequencies of prepubertal testis tumors.

Presentation and Evaluation

Most testicular tumors in children present as a painless scrotal swelling. The differential diagnosis includes epididymitis, testicular torsion, inguinal hernia, and hydrocele. On physical examination a hard mass may be palpable; however, a normal physical examination is not sufficient to exclude a tumor. Ultrasonography is helpful in distinguishing an extratesticular mass from an intratesticular mass (Fig. 11.4). A serum alpha-fetoprotein (AFP) level should also be obtained prior to treatment of a testicular mass. A serum β-HCG is obtained in pubertal boys.

The diagnosis of the type of testicular tumor is made following an inguinal orchiectomy. When the suspicion for a benign lesion is high, the tumor is excised from the testis and intraoperative histologic confirmation of a benign lesion permits testicular-sparing surgery. The type of tumor determines the need for further evaluation and follow-up.

Germ Cell Tumors

Prepubertal germ cell tumors include yolk-sac tumors, teratoma, teratocarcinoma, and seminoma. The most common germ cell tumor, as well as the most common prepubertal testicular tumor, is the yolk-sac tumor. Histologic evaluation of the yolk-sac tumor demonstrates eosinophilic PAS-positive inclusions in the cytoplasm of clear cells that consist of AFP and Schiller-Duval bodies.

AFP levels are usually elevated with yolk-sac carcinomas and may be used as a tumor marker. The half-life of AFP is about 5 days and should return to normal (<20 ng/ml) within 1 month of complete removal of the tumor. AFP levels are elevated normally in the neonate and age-specific values should be utilized. Persistent elevations of AFP postoperatively may indicate tumor metastases or recurrence. Liver dysfunction may lead to false-positive elevations of AFP.

11

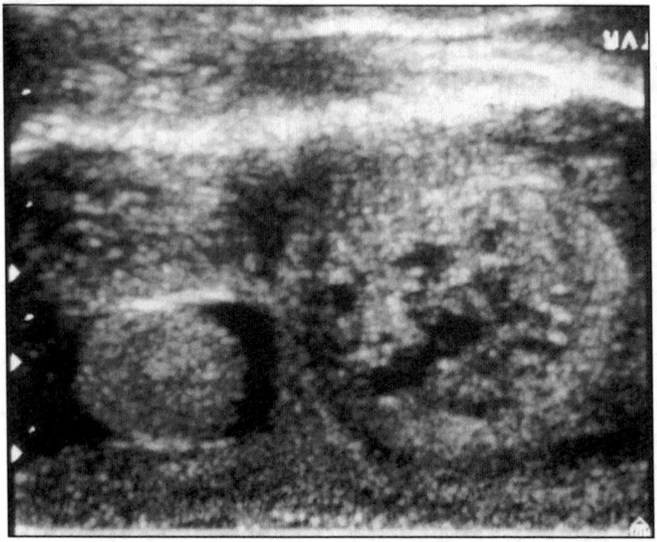

Fig. 11.4. Juvenile granulosa cell tumor. Ultrasonography of an 8-year-old demonstrating an enlarged right testicle with heterogeneous intratesticular echogenicity.

Most yolk-sac tumors occur before 2 years of age. Metastases occur predominately to the lungs (20%), and unlike the adult testicular tumor, spread via the lymphatics to the retroperitoneal nodes occurs infrequently (4-5%). Approximately 9% of patients diagnosed with yolk-sac tumor will die from their disease.

Teratoma and teratocarcinoma contain elements derived from more than one of the three germ tissues: endoderm, mesoderm, and ectoderm. These tumors are often cystic and tissues such as skin, hair, bone, and even teeth may be present. Despite containing areas of differentiation with a malignant appearance, teratomas are consistently benign in children less than 2 years of age, and may be treated by orchiectomy or testicular-sparing surgery. In the older patient with a testicular teratocarcinoma, multi-modal chemotherapy with cisplatin, bleomycin, and vinblastine results in a cure rate of 80% in those with metastatic disease. Seminoma is rare before puberty.

Gonadal Stromal Tumors

Gonadal stromal tumors include Leydig cell tumors, Sertoli cell tumors, and intermediate forms including the juvenile granulosa cell tumor. Leydig cell tumors are the most common gonadal stromal tumor in children and adults. They occur most frequently in boys 4-5 years old and synthesis of testosterone may produce precocious puberty and gynecomastia. Leydig cell tumors must also be differentiated from hyperplastic nodules that develop in boys with poorly controlled congenital adrenal hyperplasia (CAH).

Sertoli cell tumors are the second most common gonadal stromal tumors. These tumors tend to appear as a painless mass in a boy less than 6 months of age and produce no endocrinologic effects. Both Leydig cell and Sertoli cell tumors are usually benign and can be treated with local excision with testis preservation.

Although the juvenile granulosa cell tumor accounts for only 15% of all gonadal stromal tumors and 1% of all prepubertal testis tumors, it accounts for 27% of neonatal testicular tumors. These tumors tend to be nodular with variably sized, thin-walled cysts. Although there have been no reports of metastatic disease in neonates with this tumor, 10% of all granulosa cell tumors develop metastases to the nodes, liver, or lungs. The recommended treatment includes inguinal orchiectomy, periodic postoperative chest x-ray, and retroperitoneal imaging for one year.

Gonadoblastoma

Gonadoblastoma occurs in association with intersex disorders and 80% are phenotypic females with intra-abdominal testes or streak gonads. The putative gonadoblastoma gene is on the Y chromosome and the tumor almost always develops in a child whose karyotype contains a Y chromosome. The streak gonads in patients with mixed gonadal dysgenesis often develop gonadoblastomas. The peak incidence occurs at puberty accounting for current recommendations for early gonadectomy in patients at risk for gonadoblastoma. Metastatic spread of a gonadoblastoma is uncommon. These tumors may produce elevated serum levels of β-HCG. Treatment of gonadoblastoma involves removal of the gonad.

Staging

The staging for testicular tumors is demonstrated in Table 11.11.

Treatment and Outcomes

The treatment for yolk-sac tumor is inguinal orchiectomy and close surveillance. Follow-up should include monthly serum AFP levels and chest x-ray every 3 months for the first year, followed by serum AFP levels every 2 months for a second year. Routine imaging by CT or MRI of the retroperitoneum has also been recommended. Since spread to the retroperitoneal lymph nodes occurs infrequently, a routine prophylactic node dissection is not performed. Chemotherapy is given for patients with radiographic evidence of retroperitoneal disease or persistently elevated serum AFP. The use of combination chemotherapy with vincristine, actinomycin D, and cyclophosphamide, with or without doxorubicin, has been effective treatment for metastatic disease with overall survival approaching 90%.

Chemotherapy is recommended for all patients with yolk-sac tumors and stage II disease. Boys with persistent elevation of tumor markers after chemotherapy may subsequently require a retroperitoneal lymph node dissection. Boys with stage III or

11

Table 11.11. Prepubertal testicular staging

Stage I:	Tumor confined to testis and completely resected, tumor markers normalize
Stage II:	Transscrotal orchiectomy; microscopic disease in scrotum or high in spermatic cord, retroperitoneal node involvement (<2 cm); persistently abnormal tumor markers
Stage II:	Retroperitoneal lymph nodes involved (> 2 cm)
Stage IV:	Distant metastases

IV germ cell tumors are treated with chemotherapy. If elevated markers or retroperitoneal disease persists, then a biopsy or resection of residual tumor is undertaken.

Prepubertal testicular teratoma, epidermoid cyst, and stage II or higher teratocarcinomas will require treatment with cis-platin, bleomycin, and vinblastine. The recommended treatment of the juvenile granulosa cell tumor includes inguinal orchiectomy with periodic postoperative chest x-ray and retroperitoneal imaging for 1 year. Treatment of gonadoblastoma involves removal of the gonad.

Tumors in the Undescended Testis

Ten percent of germinal tumors of the testis develop in an undescended testis. The risk of malignancy in an undescended testis is 35 times greater than normal. An untreated intra-abdominal testis has a 5% risk of developing a tumor compared to a 1% risk for a tumor within the inguinal canal. With bilateral cryptorchidism, there is a 25% incidence of bilateral tumors. An increased risk of tumor development also occurs in the contralateral normally descended testis. The most common tumor that arises in an undescended testis is a seminoma, and in the testis that has been brought down into the scrotum, the most common is an embryonal carcinoma. The average age of tumor development in a formerly undescended testis is 40 years, so continued awareness of the tumor risk is emphasized.

Prepubertal orchidopexy may reduce the risk of testis cancer. The relative risk of testis cancer in a series of 1,075 boys who underwent orchidopexy in Great Britain has been reported to be 7.5 times greater than normal. The age of orchidopexy in this series did not appear to affect this risk, although few patients underwent orchidopexy prior to 5 years of age. Postpubertal orchidopexy does not diminish the risk of malignancy. Routine self-examination after puberty is appropriate for all patients with a history of an undescended testicle.

Paratesticular Rhabdomyosarcoma

Presentation and Evaluation

Seven to ten percent of rhabdomyosarcomas arise from the distal spermatic cord and present as a scrotal mass. The tumor may invade the testis, epididymis, or surrounding envelopes and be associated with a hydrocele. The initial investigation should be an ultrasound to further evaluate the mass. Benign scrotal tumors (i.e., adenomatoid, lipoma, leiomyoma, teratoma, and epidermoid cyst) should be included in the differential diagnosis. Most paratesticular rhabdomyosarcomas (97%) are embryonal and 40% of patients have involvement of the retroperitoneal lymph nodes at the time of diagnosis. Distant metastases to the lungs, cortical bone, or bone marrow occur in 20% at diagnosis. Evaluation includes a CT scan or MRI to assess the retroperitoneal nodes, as well as a chest radiograph, bone marrow aspiration, and bone scan to further assess metastases.

Treatment and Outcome

For histologic confirmation of the suspected diagnosis, the testicle should be removed using a radical inguinal orchiectomy. Transscrotal biopsies are contraindicated. Routine retroperitoneal lymph node dissection (RPLND) does not improve survival except for boys with bulky disease following chemotherapy. At present, for children under 10 years of age, the recommended treatment is a radical orchiectomy followed by retroperitoneal imaging. If there is no evidence of adenopathy, chemotherapy alone is given and long-term survival is close to 95%. For children with a T2 stage tumor or metastatic disease, treatment frequently consists of chemotherapy followed by surgical excision of any mass and additional chemotherapy or radiotherapy if indicated. Children over the age of 10 have a worse prognosis and no well-established treatment protocol exists.

Selected Readings

1. Petruzzi MJ, Green DM. Wilms' tumor. Pediatr Clin North Am 1997; 44:939-952.
2. Beckwith JB. Precursor lesions of Wilms' tumor: Clinical and biological implications. Med Pediatr Oncol 1993; 21:158-168.
3. Faria P, Beckwith JB, Mishra K et al. Focal versus diffuse anaplasia in Wilms' tumor—New definitions with prognostic significance: A report from the National Wilms' Tumor Study Group. Am J Surg Pathol 1996; 20:909-920.
4. D'Angio GJ, Breslow N, Beckwith JB et al. Treatment of Wilms' tumor. Results of the Third National Wilms' Tumor Study. Cancer 1989; 64:349-360.
5. Castleberry RP. Biology and treatment of neuroblastoma. Pediatr Clin North Am 1997; 44:919-937.
6. Crist W, Gehan EA, Ragab AH et al. The Third Intergroup Rhabdomyosarcoma Study. J Clin Oncol 1995; 13:610-630.
7. Levy DA, Kay R, Elder JS. Neonatal testis tumors: A review of the Prepubertal Testis Tumor Registry. J Urol 1994; 151:715-717.
8. Skoog SJ. Benign and malignant pediatric scrotal masses. Pediatr Clin North Am 1997; 44:1229-1250.
9. Swerdlow AJ, Higgins MC. Risk of testicular cancer in a cohort of boys with cryptorchidism. Br Med J 1997; 314:1507-1511.
10. deVries JD. Paratesticular rhabdomyosarcoma. World J Urol 1995; 13:219-225.
11. Cooper CS, Jaffe WI, Huff DS et al. The role of renal salvage procedures for bilateral Wilms' tumor: A 15-year review. J Urol 2000; 163:265-268.

11